Internet of Things

Technology, Communications and Computing

Series editors

Giancarlo Fortino, Rende (CS), Italy
Antonio Liotta, Eindhoven, The Netherlands

For further volumes:
http://www.springer.com/series/11636

Giancarlo Fortino · Paolo Trunfio
Editors

Internet of Things Based on Smart Objects

Technology, Middleware and Applications

Editors
Giancarlo Fortino
Paolo Trunfio
DIMES
University of Calabria
Rende (CS)
Italy

ISBN 978-3-319-00490-7 ISBN 978-3-319-00491-4 (eBook)
DOI 10.1007/978-3-319-00491-4
Springer Cham Heidelberg New York Dordrecht London

Library of Congress Control Number: 2014936031

Printed on acid-free paper

Springer is part of Springer Science+Business Media (www.springer.com)

Preface

The Internet of Things (IoT) usually refers to a worldwide network of interconnected heterogeneous objects (sensors, actuators, smart devices, smart objects, RFID, embedded computers, etc.) uniquely addressable, based on standard communication protocols.

Beyond such a definition, a new definition of IoT is emerging seen as a loosely coupled, decentralized system of cooperating Smart Objects (SOs). An SO is an autonomous, physical digital object augmented with sensing/actuating, processing, storing, and networking capabilities. SOs are able to sense/actuate, store, and interpret information created within themselves and around the neighboring external world where they are situated, act on their own, cooperate with each other, and exchange information with other kinds of electronic devices and human users. Their importance resides in the capabilities they have to make physical environments "smart" so as to provide novel cyber-physical services to people.

However, such SO-oriented IoT raises many "in-the-small" and "in-the-large" issues involving SO programming, IoT system architecture/middleware, and methods/methodologies for the development of SO-based applications.

This book focuses on exploring recent advances in architectures, systems, and applications for an IoT based on Smart Objects. The book specifically covers the following topics: (i) middleware for SOs; (ii) agent-oriented SOs; (iii) service-oriented SOs; (iv) Smart applications; (v) SOs indexing and discovery; (vi) IoT technologies for Smart Manufacturing; (vii) IoT technologies for Smart Grids; (viii) SOs trajectory mining for Smart City scenarios; (ix) Smart Health systems; and (x) Sensing platforms. The book is structured into ten authored chapters focused on the above-mentioned topics and provides novel and cutting-edge contributions for next-generation IoT systems. A brief introduction to the chapters is provided below.

"Middlewares for Smart Objects and Smart Environments: Overview and Comparison", by Giancarlo Fortino, Antonio Guerrieri, Wilma Russo, and Claudio Savaglio, presents an overview of middlewares for SOs and smart environments and compares them according to the most important general and specific requirements that have been identified in the literature so far. The chapter aims at providing a clear picture of the suitability of such middlewares to support the development of SO-based IoT systems.

"Mobile Agents-Based Smart Objects for the Internet of Things", by Teemu Leppänen, Jukka Riekki, Meirong Liu, Erkki Harjula, and Timo Ojala, proposes a method for the integration of mobile agents and SOs in order to facilitate cooperation and global intelligence. The chapter discusses SOs, agents, and systems requirements to enable cooperation, introduces a RESTful framework for agent creation, migration, and control, and presents an evaluation method to assist in system and agent composition design.

"Service-Oriented Middleware for the Cooperation of Smart Objects and Web Services", by Andrea Giordano and Giandomenico Spezzano, discusses how enterprise web services can be integrated with RESTful SOs by exploiting the concept of service choreography, undertaking the scalability and dynamicity issues of the IoT in order to extend the existing service composition mechanisms. The chapter shows that applications involving SO interaction can be seen as a particular case of event-driven composite services.

"CO-Based Outdoor Smart Lighting for Energy Aware Factory", by Anna Florea, Ahmed Farahat, Corina Postelnicu, Jose L. Martinez Lastra, and Francisco J. Azcondo Sánchez, describes an approach to the implementation of smart applications in a multi-purpose environment following the cooperating objects paradigm, aimed at increasing energy awareness, reducing power consumption, and enhancing user experience. As a use case, the chapter presents a smart lighting application for a multi-purpose outdoor environment.

"A Service-Oriented Discovery Framework for Cooperating Smart Objects", by Marco Lackovic and Paolo Trunfio, presents a service-oriented framework designed to support indexing, discovery, and selection of network-enabled SOs. The framework enables the dynamic discovery of distributed SOs and, specifically, their services and operations described through an ad hoc metadata model. The chapter presents the metadata model, the framework architecture and implementation, and the programming APIs.

"Smart Manufacturing Through Cloud-Based Smart Objects and SWE", by Pablo Giménez, Benjamín Molina, Carlos E. Palau, Manuel Esteve, and Jaime Calvo, discusses how IoT concepts can be applied to smart manufacturing, with smart entities that cooperate to achieve broader goals or to increase the overall knowledge in a factory through information sharing. The chapter shows that interoperability can be achieved by means of Sensor Web Enablement (SWE), while processing capabilities can be provided by virtualizing SOs on a Cloud-based datacenter.

"The Cloud of Things Empowered Smart Grid Cities", by Stamatis Karnouskos, discusses the impact that real-time monitoring and management capabilities offered by the IoT can have on the Smart Grid, and its applicability in Smart City scenarios. As an example case, the chapter highlights the efforts within the NOBEL project, which has prototyped an open service-based infrastructure for energy monitoring, management, and brokering, and points out some key aspects for the future Smart Grid City.

"Trajectory Data Analysis Over a Cloud-Based Framework for Smart City Analytics", by Eugenio Cesario, Carmela Comito, and Domenico Talia, proposes a methodology and a Cloud-based framework for trajectory pattern mining, which can be used for analyzing the trajectories of SOs in large-scale environments, particularly in Smart City scenarios. The chapter provides an experimental evaluation showing that trajectory pattern mining can take advantage from a Cloud-based parallel execution environment.

"People-Centric Service for mHealth of Wheelchair Users in Smart Cities", by Lin Yang, Wenfeng Li, Yanhong Ge, Xiuwen Fu, Raffaele Gravina, and Giancarlo Fortino, presents a real-time health-driven model for a people-centric healthcare context, introduces a social-aware architecture to support SOs mapping to online social networks, then discusses discovering and interacting with shared SOs in a virtual community. The chapter presents also a prototype for validating the proposed model.

Finally, "Experiments With a Sensing Platform for High Visibility of the Data Center", by João Loureiro, Nuno Pereira, Pedro Santos, and Eduardo Tovar, presents a hardware sensing platform for collecting physical parameters in a data center, which can serve as an enabler to optimize energy consumption. The chapter includes an analysis of the delay to obtain data from sensor networks under different data center topologies, and discusses some capabilities of the system in a real deployment.

We would like to thank all the book contributors, the anonymous reviewers, and Christoph Baumann from Springer for his support and work during the publication process.

Rende, Italy

Giancarlo Fortino
Paolo Trunfio

Contents

Middlewares for Smart Objects and Smart Environments: Overview and Comparison 1
Giancarlo Fortino, Antonio Guerrieri, Wilma Russo and Claudio Savaglio

Mobile Agents-Based Smart Objects for the Internet of Things 29
Teemu Leppänen, Jukka Riekki, Meirong Liu, Erkki Harjula and Timo Ojala

Service-Oriented Middleware for the Cooperation of Smart Objects and Web Services 49
Andrea Giordano and Giandomenico Spezzano

CO-Based Outdoor Smart Lighting for Energy Aware Factory 69
Anna Florea, Ahmed Farahat, Corina Postelnicu, Jose L. Martinez Lastra and Francisco J. Azcondo Sánchez

A Service-Oriented Discovery Framework for Cooperating Smart Objects 85
Marco Lackovic and Paolo Trunfio

Smart Manufacturing Through Cloud-Based Smart Objects and SWE 107
Pablo Giménez, Benjamín Molina, Carlos E. Palau, Manuel Esteve and Jaime Calvo

The Cloud of Things Empowered Smart Grid Cities 129
Stamatis Karnouskos

Trajectory Data Analysis Over a Cloud-Based Framework for Smart City Analytics 143
Eugenio Cesario, Carmela Comito and Domenico Talia

People-Centric Service for mHealth of Wheelchair Users in Smart Cities 163
Lin Yang, Wenfeng Li, Yanhong Ge, Xiuwen Fu, Raffaele Gravina and Giancarlo Fortino

Experiments with a Sensing Platform for High Visibility of the Data Center 181
João Loureiro, Nuno Pereira, Pedro Santos and Eduardo Tovar

Middlewares for Smart Objects and Smart Environments: Overview and Comparison

Giancarlo Fortino, Antonio Guerrieri, Wilma Russo and Claudio Savaglio

Abstract In the last few years, the Internet of Things (IoT) is gaining more and more attention both in the academic and in the industrial worlds. IoT is a concept describing a vision in which everyday objects will be connected to the Internet, will be identified, and will, possibly, communicate with other devices. These objects are typically referred as "smart objects", which can be defined as real artifacts augmented with computing, communication, sensing/actuation and storing functionalities. Their importance resides in the capabilities they have to make physical environments "smart" so as to provide novel cyberphysical services to people. In the last years, several middlewares for SOs were proposed. Middlewares, widely used in conventional distributed systems, are fundamental tools for the design and implementation of smart objects as well as of smart environment applications. They provide general and specific abstractions (e.g. object computation model, inter-object communication, sensory/actuation interfaces, discovery service, knowledge management) through which smart objects and their related applications can be easily built up. In this chapter, we present an overview of middlewares for smart objects and smart environments and compare them according to the most important general and specific requirements that have been identified in the literature so far. Moreover, such middlewares are also compared according to a feature-oriented framework to better highlight their distinctive properties. The comparison therefore provides a

G. Fortino (✉)
DIMES, University of Calabria, Rende (CS), Italy
e-mail: g.fortino@unical.it

A. Guerrieri · W. Russo · C. Savaglio
Department of Computer Engineering, Modelling, Electronics and Systems,
University of Calabria, Calabria, Italy
e-mail: aguerrieri@deis.unical.it

W. Russo
e-mail: w.russo@unical.it

C. Savaglio
e-mail: csavaglio@dimes.unical.it

G. Fortino and P. Trunfio (eds.), *Internet of Things Based on Smart Objects*,
Internet of Things, DOI: 10.1007/978-3-319-00491-4_1,

clear picture about the suitability of such middlewares to support the development of SO-based IoT systems. Finally, the chapter will briefly discuss on-going challenges in this research area.

1 Introduction

In the 1999 Kevin Ashton [1] introduced a novel paradigm named Internet of Things (IoT) [2]. The basic idea of IoT is "*the pervasive presence around us of a variety of things or objects which, through unique addressing schemes, are able to interact with each other and cooperate with their neighbors to reach common goals*" [3]. The research in this context has done several steps in the direction of a connected and collaborating world and the *things* of the IoT are starting to become smarter thanks to their augmentation with new, small, and pervasive technologies [4].

Smart Objects (SOs), which are important components of the IoT, are everyday objects that are equipped with hardware components such as a radio for communication, a CPU to process tasks, sensors/actuators to be conscious of the world in which they are situated and to control it at a given instant. Several definitions of SO are available in the literature. In [5] a smart object "*is an everyday artefact augmented with computing and communication, enabling it to establish and exchange information about itself with other digital artefacts and/or computer applications*". This definition lacks of the interaction that humans can have with SOs and that is introduced in [6], where the author proposes a "*continuous interplay between objects and people*".

Developing SOs raises many issues mostly regarding the communication among SOs, the interface with sensors/actuators, the proactivity, the knowledge management, and the distributed computation. In order to facilitate the development process, research is focused on defining new frameworks/middlewares for the rapid prototyping of SOs. These middlewares commonly provide a well defined platform and an API through which new SOs can be programmed and deployed. These SOs are able to collaborate and to support Ambient Intelligence [7] in order to build so called *Smart Environments* (SEs) [8].

The concept of SE appeared prior to the IoT. An SE can be defined as a cyber-physical environment augmented with a collection of embedded systems elaborating heterogeneous data and interacting with people. In the literature many different definitions of SE do exist, among them: (i) "*a small world, where all kinds of smart devices are continuously working to make inhabitants' lives more comfortable*" [8] and (ii) "*an environment that is able to acquire and apply knowledge about the environment and its inhabitants in order to improve their experience in that environment*" [9]. Several middlewares for the rapid prototyping of SEs have been also proposed so far [10].

This chapter aims at providing a framed overview about the main middlewares both in the SO and SE research areas. In particular, it proposes a qualitative comparison framework for such middlewares on the basis of the requirements identified in the surveyed works. Moreover, middlewares are compared with respect to several important features extrapolated from the overviewed state-of-the-art.

The chapter is organized as follows. Section 2 introduces the requirements that both SE and SO middlewares should provide. Section 3 overviews the most important middlewares for SE and SO development. In Sect. 4, starting from the identified requirements, a comparison among the overviewed work is discussed. Moreover, some important common features of the overviewed middlewares are identified and a comparison based on these characteristics is performed. Finally, conclusions are drawn and some future challenges in the IoT middleware area are discussed.

2 Middleware Requirements

The objective of this section is to analyze the requirements for both SE and SO middlewares. Such requirements will be then used to compare the most important currently available middlewares that will be systematically overviewed in the Sect. 3.

With regards to the SE middlewares, in [11] the following five general requirements have been identified:

SE_Req_1: *"Abstraction over heterogeneous input and output hardware devices"*.
Input/Output devices are usually heterogeneous so it is hard, sometimes impossible, to couple them or make them interact. Abstractions are needed to virtualize them and let them be used as they were homogeneous by following a kind of "plug-and-play" paradigm.

SE_Req_2: *"Abstraction over hardware and software interfaces"*.
Interfaces, both hardware and software, are heterogeneous too. So they need to be made generic and standardized through higher level mechanisms so that their use will be straightforward. Thus, hardware and software components based on such high-level interfaces (e.g. service oriented interfaces) will be able to seamlessly interact.

SE_Req_3: *"Abstraction over data streams (continuous or discrete data or events) and data types"*.
Different hardware and software components (e.g. sensors, devices, smart objects, mobile apps) usually produce data according to different modalities, formats and types. Thus, abstractions are needed to formalize data streams generated by such components. Both continuous data stream, discrete data and sporadic events should be defined under a common framework (e.g. through the use of XML-like languages). Moreover, the representation of data types needs to be standardized. This would allow interoperability in data exchange among heterogeneous components.

SE_Req_4: *"Abstraction over physicality (location, context)"*.
Objects in smart environments and the smart environments themselves are usually situated. This implies that they have static or dynamic locations and refer to one or multiple contexts during their lifecycle. Abstractions are therefore needed to capture and provide the concepts of location and

context as they are useful in the design and implementation of smart environment applications.

SE_Req_5: *"Abstraction over the development process"*.
To analyze, design and implement smart environments, suitable methods and tools need to be defined that are able to effectively model them by using high-level abstractions and fully support their implementation and deployment.

While the aforementioned requirements can be exploited for any smart environments (including those based on smart objects), in [12] authors defined specific requirements for smart object middlewares. In the following, they will be presented and discussed.

SO_Req_1: *"Heterogeneity and Application Development"*
Applications that use smart objects should be programmed independently from SOs produced by specific SO vendors. For instance, if an application is based on a "smart chair", it should be able to use smart chairs built by different SO vendors. Moreover, applications should be able to exploit SOs that will be built in the future. This implies to employ a standardized approach or, if not applicable as de-iure standardization is a very long process, to exploit software layering-based (dynamic) adaptation techniques between application and smart object levels.

SO_Req_2: *"Augmentation Variation of Smart Objects"*
SOs provide a set of services that can vary in number and types both among different SOs and similar SOs. In particular, different SOs can provide the same services whereas two similar SOs can provide different services. Thus, SOs cannot be crisply classified only by the object type. Moreover, they are unlike to provide standard interfaces. Augmentation variation of SOs is an important requirement as it defines how SOs can variate their augmentation by providing diversified services that can change during their lifecycle. This implies to design not only methods to dynamically add/modify/remove SO services but also how they will be actually furnished.

SO_Req_3: *"Management of Smart Object"*
An effective management of SOs is crucial in IoT applications where tons of distributed SOs could potentially interact with each other and/or be used to fulfill a final goal. Applications and SOs should be therefore able to dinamically adapt as SOs could continuously change for different purposes: augmentation variation, mobility, failures, etc. Thus, the matching among SO services and application requirements should be done often at run-time. Discovery services are therefore strategic in such a dynamic context to look up and retrieve SOs according to their static and dynamic properties.

SO_Req_4: *"Evolution of Smart Object Systems"*
Applications and SOs should be simply and rapidily prototyped and upgraded through proper programming abstractions. The evolution can be driven by programming, by learning, or by both. In particular,

evolution by learning is usually based on smart self-evolving components (application-level components and smart objects) able to self-drive their evolution on the basis of some learning model (e.g. software agents [13]).

3 Smart Environment and Smart Object Middlewares: An Overview

In this section, middlewares for the rapid prototyping of intelligent environments will be presented and, for each of them, the reference project, objectives, main features, and some applications will be described.

In particular, Sect. 3.1 introduces SE middlewares that were proposed in a context different from the IoT, as well as SE middlewares that allow to develop, as part of a SE, smart devices that exhibit some of the characteristics typical of SOs. Middlewares specifically conceived for SO development and contextualized in the IoT will be addressed in Sect. 3.2.

3.1 Middleware for Smart Environments

In this section, the main SE middlewares will be presented (ROS [11, 14], iRoom [15, 16], Aura [17–19], Context Toolkit [20, 21], JCAF [22, 23]), as well as some middlewares (Gaia [24–26], Ambient Agoras [27, 28]) that are particularly interesting since they make available some abstractions (e.g. *active spaces* and *electronic walls*) allowing the augmentation of devices.

3.1.1 ROS

Starting from the assertion "an intelligent environment is very similar to a static, non-removable robot", in [11] authors present an adaptation of middleware systems, thought for the robotic domain, to the development of smart environment: Player/Stage [29] and its evolution ROS,[1] namely Robot Operating System [14].

The main goals of ROS, summarized in [14], are the following: Peer-to-peer, Tools-based, Multi-lingual, Thin, Free and Open-Source. These goals represent the challenges encountered when developing large-scale service robots as part of the STAIR project[2] at Stanford University and the Personal Robots Program[3] at Willow Garage.

---

[1] https://vmi.lmt.ei.tum.de/ros/.

[2] http://stair.stanford.edu.

[3] http://pr.willowgarage.com.

The fundamental concepts of the ROS implementation are *nodes*, *messages*, *topics*, and *services*. Nodes are processes that perform computation and communicate with each other by passing messages. A message is a strictly typed data structure. A node sends a message by publishing it to a given topic, which is simply a string. A node that is interested to particular data will subscribe to the appropriate topic. In addition, the services support direct communication (e.g. for synchronous transactions) defined as a string name and a pair of strictly typed messages: one for the request and one for the response.

ROS has been used to implement the "*Cognitive Office*", an intelligent environment obtained by the transformation of a normal office room augmented with heterogeneous sensors and actuators managed by the ROS middleware.

3.1.2 iRoom

The Intelligent Room (iRoom) Project [15, 16], now part of the *aire* project,[4] starts from the idea to create ambients in which distributed computation adapts to environment and people just augmenting the environment with minimal *decorations* (such as camera or microphones).

The aim of the iRoom system is to track people in a room, understand what they do or say, give them news and/or suggestions, and interact with them.

The software controlling the iRoom environment is divided in three conceptual layers: the lowest one provides elements to interact with the room, the middle one wraps all the previous interaction elements in a uniform agent interface, the highest one contains agents implementing high level applications.

Agents speak to one another to control and coordinate tasks using *Metaglue* and *Hyperglue* [30], which provide simple, robust, fault tolerant communication mechanisms to software agents. Agents are implemented using SodaBot [31], an environment for creating and using distributed software agents.

The most important applications implemented in the Intelligent Room environment are related to the transformation of a room into a command center for disaster relief, to make interactive a space for virtual tours of the MIT Artificial Intelligence Laboratory. Other applications help users finding data on the Internet and organizing information about meetings.

3.1.3 Aura

Project Aura[5] [17–19] was designed with the aim to provide each user of the framework with an invisible *halo* of computing and information services that persists regardless of location.

---

[4] http://aire.csail.mit.edu/.

[5] http://www.cs.cmu.edu/~aura/.

It allows a user to transparently preserve continuity in his/her work when moving between different locations (tailoring the user's tasks to the resources in the specific environments), and adapts the computation of a particular environment in the presence of dynamic resource variability.

The Aura architecture is composed of *user tasks* (expressed as collections of services), *the task manager* which manages the user tasks, a *context observer* which allows to execute a task in the target environment, an *environment manager* that constantly monitors resources. The services needed to support a user task are specifically implemented on the target platform.

To allow different Aura environments to access user's data, the framework uses an extended version of Coda [32] which provides nomadic file access and data staging.

Several applications have been built on top of Aura with the goal of supporting on campus collaboration. An example is the *Portable Help Desk* that allows a user to know the location of teammates on the campus as well as information about them. Another application implemented for the campus is *Idealink* which provides users with features that let them communicate their ideas to others via a shared distributed blackboard.

3.1.4 Context Toolkit

The Context Toolkit project[6] [20, 21] started from the idea that the context can improve the experience in the interaction between humans and computers.

Context Toolkit is a framework for context-sensing and context-aware application development. In particular, there are three categories of features that a context-aware application can support: *presentation* of information and services to users; automatic *execution* of a service for a user; and *tagging* of context to information to support later retrieval.

The Context Toolkit provides three important abstractions: *widgets*, *interpreters*, and *aggregators*. The *context widget* [33] acquires the context and makes it available to applications in a generic manner (applications don't have to worry about how the context was sensed). Applications access the context from widgets using poll and subscribe methods. The *context interpreter* uses historical context informations to predict the future actions or intentions of users. The *context aggregator* aggregates contexts from single entities to characterize the situation of that entities.

Many applications have been built on top of Context Toolkit: *In/Out Board*, that allows to trace people entrance/exit from a research lab; *Context-Aware Mailing List*, a mailing list service that only sends emails to who is actually in a building; *Dynamic Ubiquitous Mobile Meeting BOard* (*DUMMBO*), an instrumented digitizing whiteboard that supports the capture and access of informal and spontaneous meetings; *Intercoms*, intended to facilitate conversations between occupants distributed throughout the home; *Conference Assistant*, a context-aware application that assists conference attendees.

---

[6] http://www.cs.cmu.edu/~anind/context.html.

3.1.5 JCAF

The Java Context-Awareness Framework, or JCAF[7] [22, 23], a context-awareness infrastructure and programming API, aims to make available for programmers a Java-based, lightweight and robust framework, easily extensible to support the creation of specific context-aware applications.

The main components of the framework are: the *Context Services*, that are responsible for handling context in a specific environment; the *Context Clients*, that are the context-aware applications using the JCAF infrastructure by accessing one or more context services; the *Context Monitor* and *Context Actuator* that allow, respectively, to monitor the environment acquiring context information and to affect or change the context.

JCAF was designed to be highly modifiable and extensible at runtime, then services, monitors, actuators and clients can be added to the JCAF runtime infrastructure while running.

The most important applications developed on top of JCAF are the following: the *Proximity-Based User Authentication*, that enables a user to log into a computer by approaching it; the *Context-Aware Hospital Bed* [34], a hospital bed that adjusts itself and reacts according to entities in its physical environment; the *AWARE Framework*, a system that distributes context information about users, thereby facilitating a social, peripheral awareness, which helps users to coordinate their cooperation.

3.1.6 Gaia

Gaia[8] [24–26] is a distributed middleware infrastructure capable of managing resources contained in physical spaces, called *active spaces* which simplify the development of portable applications. The framework provides mobility, adaptation, and dynamic binding enabling applications component-based, distributed and mobile. Indeed, applications can be dynamically partitioned, mapped to a variety of devices, customized on the space context, bounded to users, and moved across different spaces.

Gaia encapsulates the heterogeneity of active spaces, and presents them as a programmable environment, instead of a collection of individual and disconnected heterogeneous devices.

The system is built as a distributed object system and its three major building blocks are: the *Gaia Kernel*, the *Gaia Application Framework*, and the *Applications*. The *Gaia Kernel* contains a management and deployment system for distributed objects and an interrelated set of basic services that are used by all the applications. Gaia's *Applications* use a set of component building blocks, namely the *Gaia Application Framework*, to support applications that execute within an active space. The Gaia Application Framework allows Applications to be partitioned among a group of

[7] http://www.daimi.au.dk/~bardram/jcaf/index.html.

[8] http://gaia.cs.uiuc.edu/index.html.

coordinated devices, receive input events from different devices, present their state using different types of devices, and adapt to changes in the environment.

Gaia uses a high level scripting language, called LuaOrb [35], to program and configure active spaces and to coordinate the active entities they contain. LuaOrb is based on the interpreted language Lua [36], which simplifies management and configuration tasks and allows for rapid prototyping and testing. The interpreter for Lua is fast and has a small memory footprint, which makes it suitable for resource-constrained devices.

Specifically for Gaia a *Presentation Manager* was developed [25]. This application allows to present slideshows in multiple displays simultaneously. The Presentation Manager supports moving and duplicating slides to different displays during the presentation, and allows moving and duplicating the input sensor that controls the presentation to different devices.

3.1.7 Ambient Agoras

Ambient Agoras[9] [27, 28] is part of the European Commission funded research initiative "*The Disappearing Computer*". Ambient Agoras (from the Greek *agora*, a marketplace) wants to "turn everyday places into social marketplaces of ideas and information where people could meet and interact". In particular, the project addresses the office environment and wants to augment it by developing a wide range of smart artifacts and their corresponding software that, together, provide users with smart services.

The aims of the project were to support informal communication in organizations and to combine static artifacts integrated in the environment with mobile devices carried by people.

Seen that *ambient technology* best supports informal communication, *ambient displays* were chosen to interact with people. So, a *Hello.Wall* was developed for the Ambient Agoras environment [37]. The *Hello.Wall* provides awareness and notifications to people passing by or watching it. Different light patterns correspond to different types of information. People are known by the *Hello.Wall* thanks to mobile *ViewPort* devices provided with WLAN, RFID tags, and RFID reader to talk with other artifacts, to sense them and to be sensed.

3.2 *Middleware for Smart Objects*

This section aims to describe the former developed prototypical SO middlewares (Voyager [38, 39] and Smart-Its [40, 41]), where the concept of SO-based IoT is not well defined yet, and then to overview the SO middlewares specifically defined in the IoT context: UbiComp [42], FedNet [12, 43], Smart Products [44], and ACOSO [45–47].

[9] http://www.ambient-agoras.smart-future.net/.

3.2.1 Voyager

Authors in [38] and [39] propose Voyager,[10] a framework that supports the implementation of *ambient dialogue applications* to implement smart environments. The framework requires that every user carries with him a miniaturized pocket-size portable processing unit running user-chosen applications. Those applications are capable of dynamically detecting the presence of devices within the surrounding environment, and using those on-the-fly (realizing meaningful interaction with the user). To this end, Voyager provides the user with applications that: (i) use short-range technology (Bluetooth); (ii) discover devices in the neighborhood; (iii) communicate through a standard agreed communication protocol to query and control the services of ambient devices; (iv) use specifically implemented user API.

Voyager consists of three logical components: (i) the *client development library*, which supports the implementation of interactive applications; (ii) the *device development library*, which supports the implementation of the software to be embedded in the ambient devices; and (iii) the *standard device library*, which offers a set of implemented queries and control software components.

The implementation of Voyager has been carried out in the context of the 2WEAR project [48, 49], shared by FORTH (Greece), ETHZ (Switzerland), MA Systems and Control (UK), and NOKIA/NRC (Finland). This project aims to develop and experiment the flexible and dynamic formation of a mobile personal computing system.

Due to the short wireless range related to the low cost technology exploited, 2WEAR investigated the possibility of controlling the formation of the (mobile) personal computing system (or personal area network) based on the physical proximity of devices. Essentially, the idea was to enable users to add or remove a device to/from his personal area network respectively by approaching or by distancing himself from it.

The *Brake Out* game and the *City Guide* are two significant applications developed with the Voyager development framework. The implementation of the first one as a Voyager application has resulted in a dynamically reconfigurable user interface. The implementation of the latter one allows the users to record and visualize routes on a map in realtime using GPS and storage services. The City Guide supports dynamic configuration, as ambient devices are engaged or disengaged on-the-fly by enabling either automatic or on-demand reconfiguration. In such context, a Voyager-enabled wrist watch (or smart watch) supports users in interacting with the environment.

3.2.2 Smart-Its

Smart-Its[11] [40, 41], part of the European Commission funded research initiative "*The Disappearing Computer*", is a platform specifically designed for augmentation of everyday objects. The aim of the platform is to empower objects with

---

[10] http://2wear.ics.forth.gr/.

[11] http://www.smart-its.org.

processing, context-awareness and communication. The platform is both hardware and software. The hardware part shares some similarities with Berkeley Motes but is specifically designed for adaptation to very diverse smart objects (hardware can be specifically customized depending on the specific SO). The software part offers a set of libraries to quick implement SOs awareness, actions, communication, and logics. These libraries represent the abstractions of the hardware platform components: the *Sensors/Actuators lib* provides access to sensors and actuators; the *Communication lib* allows to access the radio and radio services; the *OS Core lib* provides access to specific OS services.

Smart-Its communicate through short range radio frequencies. Smart-Its designers didn't care about routing on their nodes, but they concentrate on fast network discovery. Smart-Its data transmission is based on a stateless P2P communication protocol.

Smart-Its devices have been used in several labs across Europe. In [50] has been presented an experiment in which objects subjected to the same movements were associated. Moreover, the platform has been used to build smart furniture, smart cups (Mediacups) [5], and to augment mobile phones. In particular, Mediacup is a smart coffee cup that is composed of an ordinary coffee cup augmented with sensors, processing and communication capabilities.

3.2.3 UbiComp

UbiComp [42], part of the DAISy (Dynamic Ambient Intelligent Systems Unit) project[12] is a middleware that supports the interaction of heterogeneous smart objects and allows to combine them through an advanced GUI application [51]. UbiComp has to run on every smart object to allow to treat SOs as components of distributed applications. UbiComp is kind of extension of the work already done for GAS (Gadgetware Architctural Style) [52].

Applications in UbiComp are component-based. The most important UbiComp component is the *artifact*. Artifacts are the tangible smart objects. Usually artifacts are augmented with sensors, actuators, processing and networking units. More artifacts can be combined in an *artifact composition*. Artifacts have *properties* which represent physical characteristics, capabilities, and services. An artifact can give values of different properties (e.g. a sensor reading) at different times. The *State* of an artifact is the snap-shot of all its properties at a given time. *Plugs* are the interfaces of an artifact. Interfaces are the set of functions that a SO exposes to its surrounding and the set of functions that the SO needs to use. When two plugs are compatible, they are associated through *Synapses*.

The SO structure [53] presents at its heart the UbiComp software middleware that allows to dynamically deploy UbiComp applications and to use sensors/actuators/ wireless communication.

In order to exemplify the usability of the middleware, a smart environment composed by a smart chair, a smart lamp, a smart book, and a smart desk was implemented. In this application, if someone sits on the chair and draws it near the desk and opens

[12] http://daisy.cti.gr/.

a book on it, then the study lamp is switched on automatically. If he closes the book or stand up, then the light goes off. The realization of the explained behavior requires the combined operation of the aforementioned artifacts.

3.2.4 FedNet

In [12, 43] authors present a framework for smart objects with the aim to make SOs collaborating among them and with applications through documents and with no direct interaction. In particular, structured documents are used to externalize (“*objectify*”) application requirements (*Task abstraction*) and SO services (*Profile abstraction*) and a runtime framework, namely FedNet, provides the spontaneous and dynamic federation of them, through structural type matching of the documents.

The framework [54] allows to coordinate heterogeneous smart objects. SOs can provide different services or the same service with different granularities. SOs can evolve at runtime and are managed by the runtime framework (located between SOs and applications) so that the applications don’t need to know the access and the configuration semantics of the SOs.

The main component of the system is the *Artefact Framework* or *Smart Object Wrapper* which encapsulates a smart object so that it can be connected and used by other smart objects and applications. It is composed by a generic core and by additional augmented features that can be added as a plug-in into the core. Each augmented feature is called *Profile*. Profiles are artefact independent. Applications in FedNet are exposed as a collection of implementation independent functional *Tasks*. FedNet maps the services requested by the tasks to the corresponding service providers (Profiles). The *FedNet Runtime Infrastructure* provides the runtime association among tasks and artefacts by using only their description documents thus externalizing smart object management and addressing heterogeneity issues away from the applications allowing developers to focus only on the application functionalities.

From a physical point of view, all these components can be distributed. To communicate, the components are forced to implement a standard RESTful (HTTP/XML) communication protocol.

Among the SOs built using FedNet, particularly interesting is the personalized peripheral information display called AwareMirror [55]. AwareMirror is an augmented mirror that presents information relevant to the person in front of it, in anticipation of the fact that users may decide to modify their behavior once an unwanted/unknown situation is understood. Moreover, a smart toothbrush [12] was developed interacting with the smart mirror.

One more case study was implemented to realize a prototype home entertainment smart object system composed by several SOs such as a smart couch, a smart lamp, a smart window, smart coffee table, a TV, an air conditioner, and bluetooth headphones. These SOs were interacting each other to form a smart environment.

3.2.5 Smart Products

Smart Products[13] [44] is a project funded as part of the Seventh Framework Programme of the EU. A Smart Product is [56] "an autonomous object which is designed for self-organized embedding into different environments and which allows for a natural product-to-human interaction."

The aim of Smart Products is to support natural interaction between objects and users and to make smart products collaborating in the environment [57]. Smart Products activity is mainly based on workflows that developers specify for each procedure they want the smart product to support.

The component-based architecture of the Smart Products platform [58] provides components for supporting the interaction between user and smart products; components for supporting the information exchange and cooperation between different smart products; components for sensing, processing and distributing context information; components for handling the knowledge of a smart product; components for storing knowledge of a smart product in a secure and distributed way.

The communication components are based on the MundoCore middleware [59], which provides peer to peer communication and discovery.

Among the applications implemented in this project, Smart Products have been used to assist users in assembly actions, in the smart car application domain (snow chain assembly, tyre change, etc.), and as smart cooking service aware of users and appliances.

3.2.6 ACOSO

The ACOSO (Agent-based COoperating Smart Objects) middleware [45–47] was developed in the context of the TETRis (TETRA Innovative Open Source Services) project, funded by the Italian Government.

The aim of the middleware is to provide the right instruments and a simple programming model to realize cooperating smart objects.

ACOSO is agent-oriented and supplies the SOs with an event-driven architecture that allows the SOs to fastly react to external stimulus. The middleware supports several communication types (message passing and publish/subscribe) and provides proactivity based on inference rules and on local and remote Knowledge Bases.

While the design of the architecture is platform independent and can be realized on several agent-based platforms (e.g. JADE [60], Jadex [61], MAPS [62]), actually, the middleware implementation relies on the JADE middleware that provides an effective agent management and infrastructure of communication. So, the SOs can be implemented as JADE or Jadex agents.

The architecture proposed for the SOs is very simple and effective and composed by a *Behavior* incorporating the object behavior, an *EventDispatcher* handling all the internal events of the object, a *Communication Management Subsystem* controlling

---

[13] http://www.smartproducts-project.eu/.

communications with other SOs and external entities, a *Device Management Subsystem* coordinating the sensor/actuator nodes of the object, and a *KB Management Subsystem* managing the knowledge base in the SO.

ACOSO is completely integrated with two frameworks realized to control wireless sensor networks, such as the Building Management Framework [63] and the SPINE framework [64].

In [46] a case study was presented referring to a smart environment composed of two cooperating SOs: a Smart Office and a Smart Body. The two SOs gather information about the environment, the body, and the interaction between the body and the environment and cooperate to support the working activity of the office user (e.g. giving suggestions about wastes in the environment that can be limited). Moreover, the Smart Office is based on a smart chair, a smart projector and a smart whiteboard.

4 Comparison and Discussion

This section presents a detailed comparison of the middlewares described in Sect. 3. In particular, Sect. 4.1 shows a comparison based on the requirements introduced in Sect. 2 while Sect. 4.2 provides a comparison based on the most important identified features which allow to characterize SE and SO middlewares.

4.1 Middleware Requirements-Based Comparison

The requirements introduced in Sect. 2 can be used to compare all the proposals analyzed in Sect. 3. In particular, both SE and SO middlewares can be successfully compared on the basis of the requirements SE_Req_{1-5}.

Specifically, all the overviewed works, with the exception of Ambient Agoras, fulfill the requirements SE_Req_1, SE_Req_2 and SE_Req_5. Regarding the requirement SE_Req_3, it is not satisfied by Smart-Its, Context Toolkit, Gaia, iRoom, and JCAF that do not provide generic data stream formalization, but specific ones. The requirement SE_Req_4 is the less considered among the examined work. In particular, UbiComp, Gaia, Smart-Its, Ambient Agoras, Voyager, Context Toolkit, Aura, and JCAF do not provide methods to abstract applications (for SE middlewares) or SOs (for SO middlewares) over the location on which they will be deployed.

The full comparison of SE and SO middlewares with respect to the general SE middleware requirements is reported in Table 1.

Moreover, a further comparison of the works analyzed in Sect. 3 is carried out with respect to the specific requirements for SO middlewares (SO_Req_{1-4}). Middlewares for SEs are also included in this comparison, see Table 2, to bring out if some of them also fulfills the specific SO middlewares requirements and can be therefore reused in the SO context.

Table 1 Comparison of the overviewed SE/SO middlewares based on the general SE middlewares requirements

	SO/SE middleware	SE Req1	SE Req2	SE Req3	SE Req4	SE Req5
ROS-player/stage	SE	Yes	Yes	Yes	Yes	Yes
iRoom	SE	Yes	Yes	No	Yes	Yes
Aura	SE	Yes	Yes	Yes	No	Yes
Context toolkit	SE	Yes	Yes	No	No	Yes
JCAF	SE	Yes	Yes	No	No	Yes
Gaia	SE	Yes	Yes	No	No	Yes
Ambient agoras	SE	N/A	N/A	N/A	No	N/A
Voyager-2WEAR	SO	Yes	Yes	Yes	No	Yes
Smart-Its	SO	Yes	Yes	No	No	Yes
UbiComp/GAS	SO	Yes	Yes	Yes	No	Yes
FedNet	SO	Yes	Yes	Yes	Yes	Yes
Smart products	SO	Yes	Yes	Yes	Yes	Yes
ACOSO	SO	Yes	Yes	Yes	Yes	Yes

Table 2 Comparison of the overviewed SE/SO middlewares based on the specific SO middlewares requirements

	SO/SE middleware	SO Req1	SO Req2	SO Req3	SO Req4
ROS-player/stage	SE	No	Yes	Yes	Yes
iRoom	SE	No	No	Yes	Yes
Aura	SE	No	Yes	No	Yes
Context toolkit	SE	No	No	No	Yes
JCAF	SE	No	No	No	Yes
Gaia	SE	Yes	No	No	Yes
Ambient agoras	SE	N/A	N/A	No	N/A
Voyager-2WEAR	SO	No	Yes	No	Yes
Smart-its	SO	Yes	No	No	Yes
UbiComp/GAS	SO	Yes	Yes	No	Yes
FedNet	SO	Yes	Yes	Yes	Yes
Smart products	SO	Yes	Yes	Yes	Yes
ACOSO	SO	Yes	Yes	Yes	Yes

As highlighted in Table 2, the requirement SO_Req_1 is the most distinctive for SO middlewares because, to fulfill it, the SO abstraction has to be provided by the middleware itself. The only SE middleware satisfying the SO_Req_1 is Gaia as it makes available a particular abstraction sharing some similarities with the SO, namely the *active space*.

With reference to the requirement SO_Req_2, all the SO middlewares, apart from Smart-Its, do not provide standard interfaces to access the SOs. Regarding the SE middlewares, Aura and ROS satisfy this requirement since they provide a flexible interface among SE components or different SEs.

Requirement SO_Req_3 is satisfied by ACOSO, FedNet, and Smart Products that implement at middleware-level a mechanism to match the application requirements with the services provided by the SOs. Among the middlewares for SEs, only iRoom and ROS provide mechanisms to match SE services and application requirements.

Requirement SO_Req_4 is satisfied by all the SO and SE middlewares since all the proposals (apart Ambient Agoras which does not provide details about such requirement) offer programming abstractions to support SO/SE implementation and enhancement. However, the evolution is only driven by programming.

4.2 Middleware Features-Oriented Comparison

This section presents a comparison of the works presented in Sect. 3, based on the most important characteristics of the middlewares for SEs and SOs. These characteristics are distinctive properties of the middlewares and have been chosen after a detailed analysis of the literature to determine a feature-oriented comparison framework among the surveyed work.

In particular, the majority of the features have been considered for both SE and SO middlewares, while some of them are specifically referred to either SE or SO middlewares.

The common features are listed in the following highlighting the reference model and the implementing technology. They are:

- *System Programming*, the programming model used by the middleware (e.g. component-based, agent-based, object-based, service-based) and the specific programming language exploited.
- *System Architecture*, the system architecture of the middleware (e.g. client/server, peer-to-peer, distributed, hierarchical) and its related technology.
- *Metadata*, the model for the representation of information about SEs or SOs and its implementing technology.
- *Discovery*, the model on which the middleware is based to discover SE elements or SOs (e.g. centralized, distributed) along with the implementation method.
- *Communication*, the communication paradigm through which the entities/components/objects in the system communicate (e.g. peer-to-peer, client/server, tuple-based, publish/subscribe) and the communication technology.
- *Knowledge Management*, if the knowledge is managed by the middleware and, if available, where it is managed (e.g. local, distributed, hybrid) and through which technology.

With reference to SE middlewares, there are the following specific features:

- *SE Proactivity*, if the SE is proactive (i.e. it is able to autonomusly fulfill goals) and, if it is, how the proactivity is made available.
- *SE Cooperation*, if the SE is cooperative (i.e. it is able to cooperate with other SE) and, if it is, how the proactivity is made available.
- *SE/Application decoupling*, if applications running atop the SE are decoupled from the SE components.

Finally, with reference to the SO middlewares the following characteristics are considered:

- *SO Architecture*, the specific architectural model defined for developing SOs (e.g. agent-oriented, object-oriented, service-oriented, component-based) and the techonology used to implement the model.
- *SO Proactivity*, whether or not SOs are proactive (i.e. it is able to autonomously fulfill goals) and, if they are, how the proactivity is made available.
- *SO Cooperation*, whether or not SOs are cooperative (i.e. it is able to cooperate with other SE) and, if they are, how the proactivity is made available.
- *SO/Application coupling*, if applications running atop SOs are decoupled from SOs.

The comparison, on the basis of the defined features, is shown in Table 3 and in Table 4 for SE and SO middlewares, respectively.

From the analysis of Table 3, three main considerations arise:

- Surveyed middlewares have a marked heterogeneity in the adoption of models in relation to all the considered features. The only point of contact seems to be the architectural model, which of course is distributed.
- A strong heterogeneity also with regard to technology is shown, although Java and CORBA are the most commonly used (in a cross to the different characteristics).
- Almost all of the surveyed SE middlewares do not make available services regarding Metadata, Discovery services (with the exception of Context Toolkit), and Knowledge Management (with the exception of Aura).

In principle, these considerations remain valid even after the analysis of SO middlewares (see Table 3), although they appear functionally more complete of the corresponding SE middlewares. In particular:

- The heterogeneity of the adopted models is still present, although to a lesser extent: distributed system architectures, component-based SO architecture, and Message Passing as communication strategy are in fact commonly used.
- C-based and Java-based applications technologies are exploited in a cross.
- Different SO middlewares take into account the Metadata but are lacking about Code Availability (with the exceptions of Smart Products and ACOSO) and SO/App decoupling (totally supported only by FedNet and partially only by ACOSO).

5 Conclusions and Future Research Challenges

This chapter has presented an overview of the state-of-the-art regarding SE and SO middlewares. In particular, the main SE and SO middlewares have been described and compared with reference to three comparison frameworks: (i) the first one is based on general SE middleware requirements that have been elicited in the literature;

Table 3 Smart environment middlewares

		ROS	iRoom	Aura	Context toolkit	JCAF	Gaia	Ambient agoras
System programming								
	Model	Component-based	Agent-based	Task-oriented	Component based	Service-based	Component based	N/A
	Technology	Multi-language (C++, Java, Python, Lisp, etc)	SodaBotL language	Java	Java	Java	Lua script language	N/A
System Architecture								
	Model	Distributed	Distributed	Distributed	Distributed	Distributed	Distributed	N/A
	Technology	C++	SodaBot	Java/RMI-CORBA	Java	Java	CORBA	N/A
SE proactivity		By programming	No	By model	By programming	No	No	N/A
SE cooperation		By programming	By model	By model	By programming	No	No	N/A
Metadata								
	Model	N/A	N/A	N/A	N/A	N/A	N/A	N/A
	Technology	N/A	N/A	N/A	N/A	N/A	N/A	N/A
Discovery								
	Model	N/A	N/A	N/A	Centralized	N/A	N/A	N/A
	Technology	N/A	N/A	N/A	HTTP server	N/A	N/A	N/A
Communication								
	Model	Topic-based publish/subscribe	RPC and publish/subscribe	Message passing	Message passing	Event-based publish/subscribe	Message passing	(Low-level) Message passing
	Technology	XML-RPC	Metaglue and hyperglue	CORBA, RPC	HTTP	Java RMI	CORBA	WIFI/RFID

continued

Table 3 continues

Knowledge Management								
	Model	N/A	N/A	Local and remote	Local	N/A	N/A	N/A
	Technology	N/A	N/A	Coda	Java	N/A	N/A	N/A
SE/App decoupling		Partial	Partial	No	No	Partial	No	No
Code availability		Yes	Yes	Yes	Yes	Yes	No	No

Table 4 Smart objects middlewares

		Voyager	Smart-Its	Ubicomp	FedNet	Smart products	ACOSO
System programming							
	Model	Component-based	Object based	Component-based	Component based	Component based	Agent-based
	Technology	C++	C	Java/J2ME	Java	Java	Java
System Architecture							
	Model	Distributed	Distributed	Distributed	Distributed	Distributed	Distributed
	Technology	C++	C	Java/J2ME	Java	Mundocore	JADE
SO architecture							
	Model	Component-based	Not specified	Component-based	Component-based	Component-based	Event-based and agent-oriented
	Technology	C++	C	Java/J2ME	Java	Java	JADE and JADEX
SO proactivity		No	By programming	By model	No	By model	By model
SO cooperation		No	By programming	By model	No	By model	By model
Metadata							
	Model	N/A	N/A	<Class, Communication, Resources>	Profile abstraction	<Service>	<Type, Device, Services, Location> model.
	Technology	N/A	N/A	XML	XML	Java objects	JSON
Discovery							
	Model	Distributed	Distributed	Distributed	Centralized	Distributed	Centralized
	Technology	Bluetooth-based	Bluetooth-based	eRDP/UDP/IP	RESTful	Mundocore	RESTful Web Services + Lucene
Communication							
	Model	Message passing	Message passing	Message passing	Message passing	Message passing	Asynchronous Message Passing + Publish/Subscribe
	Technology	L2CAP/Bluetooth	Bluetooth-based P2P stateless protocol	Java sockets	RESTful communication protocol	MundoCore	JADE ACL Messages + JADE topic-based P/S

continued

Table 4 continues

Knowledge Management							
	Model	Local	N/A	Local	N/A	Local and remote	Local (and remote)
	Technology	Control protocol/L2CAP/ bluetooth	N/A	Prolog engine	N/A	OWL, RDF/ MundoCore	JADE and JADEX
SO/App decoupling		No	No	No	Full	No	Partial
Code availability		No	No	No	No	Yes	Yes

(ii) the second one relies on specific SO middleware requirements identified in the literature; (iii) the last one is based on a set of features specifically identified on the basis of the overviewed works. Overall, the proposed comparison provides a clear picture about the characteristics of the surveyed middlewares and, specifically, their suitability in supporting the development of SO-based applications in the IoT context. In particular, ACOSO, FedNet, Ubicomp and Smart Products are fully suitable to be actually exploited, Voyager and Smart-Its are partially suitable but need to be enhanced, whereas the surveyed SE middlewares should be heavily enhanced to be practically used.

Although the overviewed SO middlewares provide effective management to develop and deploy SOs, they are still limited in the management of a huge number of cooperative SOs. According to forecasts from Cisco Systems [65], by 2020 more than 50 billion of smart objects will be connected in the IoT. It becomes clear that such a huge swarm of devices cannot be subject to a human-driven management, because of the intrinsic complexity of such a scenario and the heterogeneity of its components. In addition, several application contexts may require response times decidedly out of the reach of an operator, though very skilled. So, the entire system must be able to auto configure itself following specific and predetermined objectives and policies, as well as keep itself functioning, efficient, reliable and protected against endogenous and/or exogenous dangers or malfunctions, without the human intervention [66, 67]. In recent years, systems self-* were massively used for autonomic management of large scenarios [68], characterized however by well defined interactions. In such situations, the autonomic architecture properties can provide flexibility and adaptability in the operational management (optimization of communications, services discovery, adoption of protection policies and strategies), minimizing the resources used through sets of predetermined rules. At the same time, IoT expectations regarding the context-awareness and user-tailored content management have highlighted the need for interoperability, abstraction, collective intelligence, dynamism and experience-based learning. To cope with these requirements cognitive systems, typically agent-based, were widely employed [69, 45]. Multi Agent Systems, already used in contexts typically not extensive but characterized by an high uncertainty, prove to be the right means to infuse intelligence into smart objects and in the whole big scenario. Inspired by the human body, in which the two mechanisms operate synergistically, one could hypothesize a cognitive-autonomic management for the IoT, with the aim of making its realization as close to the theoretical expectations.

In addition to a cognitive-autonomic management of cooperative SOs, complementary approaches need to be exploited to enhance computing and memory capabilities of SOs. The Cloud Computing paradigm provides flexible, robust and powerful storage and computing resources, and enables dynamic data integration and fusion from multiple data sources [70]. Cloud Computing can therefore support SO, empowering their specific resources, and allow for the definition of new and more capable (virtual) SOs as meta-aggregation of existing real and virtual SOs [71].

Finally, to develop SO-based IoT systems, novel software engineering methodologies for extreme-scale dynamic systems need to be defined. They need to include not only a modeling language along with a well-defined process but also specific abstractions able to deal with system/component evolution that is a typical property of SO systems. Agent-oriented methodologies [72] could be exploited as basis for formalizing such an effective development methodology for SOs.

Acknowledgments This work has been partially supported by TETRis TETRA Innovative Open Source Services, funded by the Italian Government (PON 01-00451). Giancarlo Fortino is also partially funded by DICET INMOTO Organization of Cultural Heritage for Smart Tourism and REal Time Accessibility (OR.C.HE.S.T.R.A.) funded by the Italian Government (PON04a2_D).

References

1. Ashton, K.: That 'internet of things' thing, in the real world things matter more than ideas. RFID J. (2009). http://www.rfidjournal.com/articles/view?4986
2. Santucci, G.: From internet of data to internet of things. In: Proceedings of the International Conference on Future Trends of the Internet (2009)
3. Atzori, L., Iera, A., Morabito, G.: The internet of things: a survey. Comput. Netw. **54**(15), 2787–2805 (2010). http://dx.doi.org/10.1016/j.comnet.2010.05.010
4. Bandyopadhyay, D., Sen, J.: Internet of things: applications and challenges in technology and standardization. Wirel. Pers. Commun. **58**(1), 49–69 (2011). http://dx.doi.org/10.1007/s11277-011-0288-5
5. Beigl, M., Gellersen, H., Schmidt, A.: Mediacups: experience with design and use of computer-augmented everyday artefacts. Comput. Netw. **35**(4), 401–409 (2001). http://dx.doi.org/10.1016/S1389-1286(00)00180--8
6. Sterling, B.: Shaping Things. MIT Press, Cambridge (2005). http://mitpress.mit.edu/catalog/item/default.asp?ttype=2&tid=10603
7. Aarts, E., de Ruyter, B.: New research perspectives on Ambient intelligence. J. Ambient Intell. Smart Environ. **1**(1), 5–14 (2009). http://dl.acm.org/citation.cfm?id=1735821.1735822
8. Cook, D.J., Das, S.K.: How smart are our environments? an updated look at the state of the art. Pervasive Mob. Comput. **3**(2), 53–73 (2007). http://dx.doi.org/10.1016/j.pmcj.2006.12.001
9. Youngblood, G.M., Heierman, E.O., Holder, L.B., Cook, D.J.: Automation intelligence for the smart environment. In: Proceedings of the 19th International Joint Conference on Artificial Intelligence, ser. IJCAI'05. San Francisco, CA, USA, Morgan Kaufmann Publishers Inc., pp. 1513–1514 (2005). http://dl.acm.org/citation.cfm?id=1642293.1642535
10. Badica, C., Brezovan, M., Badica, A.: An overview of smart home environments: architectures, technologies and applications. In: Georgiadis, C.K., Kefalas, P., Stamatis, D. (eds.) Proceedings of the Sixth Balkan Conference in Informatics. BCI 2013, vol. 1036 (2013). http://CEUR-WS.org
11. Roalter, L., Kranz, M., Möller, A.: A middleware for intelligent environments and the internet of things. In: Proceedings of the 7th International Conference on Ubiquitous Intelligence and Computing. UIC'10, pp. 267–281. Springer, Berlin, Heidelberg (2010). http://dl.acm.org/citation.cfm?id=1929661.1929690
12. Kawsar, F., Nakajima, T.: A document centric framework for building distributed smart object systems. In: Proceedings of the 2009 IEEE International Symposium on Object/Component/Service-Oriented Real-Time Distributed Computing. ISORC '09, pp. 71–79. IEEE Computer Society, Washington, DC, USA (2009). http://dx.doi.org/10.1109/ISORC.2009.16
13. Fortino, G., Galzarano, S.: On the development of mobile agent systems for wireless sensor networks: issues and solutions. Intell. Syst. Ref. Libr. **45**, 185–215 (2013). cited By (since 1996)0. http://www.scopus.com/inward/record.url?eid=2-s2.0-84885636987&partnerID=40&md5=70b6432708416f13aba1f06fa36b1e4b

14. Quigley, M., Conley, K., Gerkey, B.P., Faust, J., Foote, T., Leibs, J., Wheeler, R., Ng, A.Y.: ROS: an open-source robot operating system. In: ICRA Workshop on Open Source Software (2009)
15. Brooks, R.A.: The intelligent room project. In: Proceedings of the 2nd International Conference on Cognitive Technology (CT '97). CT '97, pp. 271–278. IEEE Computer Society, Washington, DC, USA (1997). http://dl.acm.org/citation.cfm?id=794204.795308
16. Coen, M.: The future of human-computer interaction, or how i learned to stop worrying and love my Intelligent Room. IEEE Intell. Syst. **14**(2), 8–19 (1999)
17. Sousa, J.P., Garlan, D.: Aura: an architectural framework for user mobility in ubiquitous computing environments. In: Proceedings of the IFIP 17th World Computer Congress—TC2 Stream / 3rd IEEE/IFIP Conference on Software Architecture: System Design, Development and Maintenance. WICSA 3, pp. 29–43. Kluwer, B.V., Deventer, The Netherlands (2002). http://dl.acm.org/citation.cfm?id=646546.693943
18. Garlan, D., Siewiorek, D., Smailagic, A., Steenkiste, P.: Project aura: toward distraction-free pervasive computing. IEEE Pervasive Comput. **1**(2), 22–31 (2002). http://dx.doi.org/10.1109/MPRV.2002.1012334
19. Hengartner, U., Steenkiste, P.: Implementing access control to people location information. In: Proceedings of the Ninth ACM Symposium on Access Control Models and Technologies. SACMAT '04, pp. 11–20 . New York, NY, USA, ACM (2004). http://doi.acm.org/10.1145/990036.990039
20. Dey, A.K.: Understanding and using context. Pers. Ubiquitous Comput. **5**(1), 4–7 (2001). http://dx.doi.org/10.1007/s007790170019
21. Dey, A.K., Abowd, G.D., Salber, D.: A conceptual framework and a toolkit for supporting the rapid prototyping of context-aware applications. Hum. Comput. Interact. **16**(2), 97–166 (2001). http://dx.doi.org/10.1207/S15327051HCI16234_02
22. Bardram, J.E.: The java context awareness framework (JCAF)—a service infrastructure and programming framework for context-aware applications. In: Proceedings of the Third International Conference on Pervasive Computing, ser. PERVASIVE'05, pp. 98–115. Springer, Berlin, Heidelberg (2005). http://dx.doi.org/10.1007/11428572_7
23. Bardram, J.E., Kjær, R.E., Pedersen, M.: Context-aware user authentication—supporting proximity-based login in pervasive computing. In: Dey, A., Schmidt, A., McCarthy, J. (eds.) UbiComp 2003: Ubiquitous Computing. Lecture Notes in Computer Science, vol. 2864, pp. 107–123. Springer, Berlin (2003). http://dx.doi.org/10.1007/978-3-540-39653-6_8
24. Roman, M., Campbell, R.H.: Gaia: enabling active spaces. In: Proceedings of the 9th Workshop on ACM SIGOPS European Workshop: Beyond the PC: New Challenges for the Operating System. EW 9, pp. 229–234. New York, NY, USA, ACM (2000). http://doi.acm.org/10.1145/566726.566772
25. Romn, M., Hess, C., Cerqueira, R., Campbell, R., Nahrstedt, K.: Gaia: a middleware infrastructure to enable active spaces. IEEE Pervasive Comput. **1**, 74–83 (2002)
26. Román, M., Hess, C., Cerqueira, R., Ranganathan, A., Campbell, R.H., Nahrstedt, K.: A middleware infrastructure for active spaces. IEEE Pervasive Comput. **1**(4), 74–83 (2002). http://dx.doi.org/10.1109/MPRV.2002.1158281
27. Streitz, N., Prante, T., Röcker, C., van Alphen, D., Magerkurth, C., Stenzel, R., Plewe, D.: Ambient Displays and Mobile Devices for the Creation of Social Architectural Spaces: Supporting Informal Communication and Social Awareness in Organizations, ch. 16, pp. 387–409. Kluwer Publishers, Netherlands (2003)
28. Streitz, N.A., Röcker, C., Prante, T., van Alphen, D., Stenzel, R., Magerkurth, C.: Designing smart artifacts for smart environments. Computer **38**(3), 41–49 (2005). http://dx.doi.org/10.1109/MC.2005.92
29. Collett, T.H.J., Macdonald, B.A.: Player 2.0: toward a practical robot programming framework. In: Proceedings of the Australasian Conference on Robotics and Automation (ACRA) (2005)
30. Hanssens, N., Kulkarni, A., Tuchida, R., Horton, T.: Building agent-based intelligent workspaces. In: Proceedings of the International Conference on Internet Computing, pp. 675–681 (2002). http://dblp.uni-trier.de/db/conf/ic/ic2002-3.html#HanssensKTH02

31. Coen, M.: SodaBot: a software agent construction system. In: Proceedings of the 1994 Conference on Information and Knowledge Management Workshop on Intelligent Information Agents, MIT AI, p. 84 (1995)
32. Satyanarayanan, M., Kistler, J.J., Kumar, P., Okasaki, M.E., Siegel, E.H., Steere, D.C.: Coda: a highly available file system for a distributed workstation environment. IEEE Trans. Comput. **39**(4), 447–459 (1990). http://dx.doi.org/10.1109/12.54838
33. Schilit, B., Adams, N., Want, R.: Context-aware computing applications. In: Proceedings of the 1994 First Workshop on Mobile Computing Systems and Applications. WMCSA '94, pp. 85–90. IEEE Computer Society, Washington, DC, USA (1994). http://dx.doi.org/10.1109/WMCSA.1994.16
34. Bardram, J.E.: Applications of context-aware computing in hospital work: examples and design principles. In: Proceedings of the 2004 ACM Symposium on Applied Computing. SAC '04, pp. 1574–1579. New York, NY, USA, ACM (2004). http://doi.acm.org/10.1145/967900.968215
35. Cerqueira, R., Cassino, C., Ierusalimschy, R.: Dynamic component gluing across different componentware systems. In: Proceedings of the International Symposium on Distributed Objects and Applications. DOA '99, pp. 362–371. IEEE Computer Society, Washington, DC, USA (1999) http://dl.acm.org/citation.cfm?id=519622.790758
36. Ierusalimschy, R., de Figueiredo, L.H., Filho, W.C.: Lua—an extensible extension language. Softw. Pract. Exper. **26**(6), 635–652 (1996). http://dx.doi.org/10.1002/(SICI)1097--024X(199606)26:6<635::AID-SPE26>3.0.CO;2-P
37. Röcker, C., Prante, T., Streitz, N., van Alphen, D.: Using ambient displays and smart artefacts to support community interaction in distributed teams. In: Proceedings of the 16th Annual Conference of the Australian Computer-Human Interaction Special Interest Group (OZCHI 2004), (2004)
38. Savidis, A., Stephanidis, C.: Distributed interface bits: dynamic dialogue composition from ambient computing resources. Pers. Ubiquitous Comput. **9**(3), 142–168 (2005). http://dx.doi.org/10.1007/s00779-004-0327-2
39. Savidis, A., Stephanides, C.: Dynamic environment-adapted mobile interfaces: the voyager toolkit. In: Array (ed.) Proceedings of the 10th International Conference on Human-Computer Interaction (HCI International 2003), vol. 4, pp. 489–493. Lawrence Erlbaum Associates, New Jersey (2003)
40. Gellersen, H.W., Schmidt, A., Beigl, M.: Multi-sensor context-awareness in mobile devices and smart artifacts. Mob. Netw. Appl. **7**(5), 341–351 (2002). http://dx.doi.org/10.1023/A:1016587515822
41. Beigl, M., Gellersen, H.W.: Smart-its: an embedded platform for smart objects. In: Proceedings of the Smart Object Conference (SOC 2003), pp. 15–17 (2003)
42. Goumopoulos, C., Kameas, A.: Smart objects as components of UbiComp applications. In: International Journal of Multimedia and Ubiquitous Engineering. Special Issue on Smart Object Systems, vol. 4, no. 3, pp. 1–20. sERSC Press, Springer (2009).
43. Kawsar, F., Nakajima, T., Park, J.H., Yeo, S.S.: Design and implementation of a framework for building distributed smart object systems. J. Supercomput. **54**(1), 4–28 (2010). http://dx.doi.org/10.1007/s11227-009-0323-4
44. Mühlhäuser, M.: Smart products: an introduction. In: Constructing Ambient Intelligence: Am I 2007 Workshops, Springer, pp. 158–164 (2007)
45. Fortino, G., Guerrieri, A., Russo, W.: Agent-oriented smart objects development. In: Proceedings of the 2012 IEEE 16th International Conference on Computer Supported Cooperative Work in Design (CSCWD), pp. 907–912 (2012)
46. Fortino, G., Guerrieri, A., Lacopo, M., Lucia, M., Russo, W.: An agent-based middleware for cooperating smart objects. In: Corchado, J., Bajo, J., Kozlak, J., Pawlewski, P., Molina, J., Julian, V., Silveira, R., Unland, R., Giroux, S. (eds.) Highlights on Practical Applications of Agents and Multi-Agent Systems. Communications in Computer and Information Science, vol. 365, pp. 387–398. Springer, Berlin, (2013). http://dx.doi.org/10.1007/978-3-642-38061-7_36

47. Fortino, G., Lackovic, M., Russo, W., Trunfio, P.: A discovery service for smart objects over an agent-based middleware. In: Pathan, M., Wei, G., Fortino, G. (eds.) Proceedings of the 6th International Conference on Internet and Distributed Computing Systems. IDCS 2013, pp. 1–13, vol. 8223. Springer, Heidelberg, Berlin (2013)
48. Lalis, S., Savidis, A., Stephanides, C.: Supporting distributed user interfaces in mobile and wearable device ensembles: the 2WEAR experience. In: Proceedings of the Ensembles of On-Body Devices workshop at Mobile HCI 2010 (2010)
49. Lalis, S., Savidis, A., Karypidis, A., Gutknecht, J.: Towards dynamic and cooperative multi-device personal computing. In: Streitz, N., Kameas, A., Mavrommati, I. (eds.) The disappearing computer, pp. 182–204. Springer, Berlin, (2007). http://dl.acm.org/citation.cfm?id=1768246.1768258
50. Holmquist, L.E., Mattern, F., Schiele, B., Alahuhta, P., Beigl, M., Gellersen, H.W.: Smart-its friends: a technique for users to easily establish connections between smart artefacts. In: Proceedings of the 3rd International Conference on Ubiquitous Computing. UbiComp '01, pp. 116–122. Springer-Verlag, London, UK (2001). http://dl.acm.org/citation.cfm?id=647987.741340
51. Mavrommati, I., Kameas, A., Markopoulos, P.: An editing tool that manages device associations in an in-home environment. Pers. Ubiquitous Comput. **8**(3–4), 255–263 (2004). http://dx.doi.org/10.1007/s00779-004-0286-7
52. Kameas, A., Bellis, S., Mavrommati, I., Delaney, K., Colley, M., Pounds-Cornish, A.: An architecture that treats everyday objects as communicating tangible components. In: Proceedings of the First IEEE International Conference on Pervasive Computing and Communications. PERCOM '03. IEEE Computer Society, Washington, DC, USA, pp. 115–122 (2003). http://dl.acm.org/citation.cfm?id=826025.826369
53. Drossos, N., Kameas, A.: Building composeable smart objects. In: Proceedings of the 1st International Workshop on Design and Integration Principles for Smart Objects. DIPSO 2007, Innsbruck, Austria (2007)
54. Kawsar, F., Kortuem, G., Altakrouri, B.: Supporting interaction with the internet of things across objects, time and space. In: Proceedings of the 2nd International Internet of Things Conference. IEEE, pp. 1–8 (2010). iSBN: 978-1-4244-7413-4 IEEE Catalog Number: CFP1014K-ART
55. Fujinami, K., Kawsar, F.: Embedding context-awareness into a daily object for improved information awareness: A Case Study Using a Mirror. In N. Chong, F. Mastrogiovanni (eds.) Handbook of Research on Ambient Intelligence and Smart Environments: Trends and Perspectives, pp. 56–77. Information Science Reference, Hershey, PA (2011). doi:10.4018/978-1-61692-857-5.ch004
56. Sabou, M., Kantorovitch, J., Nikolov, A., Tokmakoff, A., Zhou, X., Motta, E.: Position paper on realizing smart products: challenges for Semantic Web technologies. In: Proceedings of the 2nd International Workshop on Semantic Sensor Networks 2009 (SSN09) at ISWC 2009, vol. 522, pp. 135–147 (2009). http://oro.open.ac.uk/23449/
57. Ständer, M.: Towards interactionflows for smart products. In: Proceedings of the 2010 ACM Symposium on Applied Computing. SAC '10. ACM, New York, NY, USA (2010). http://doi.acm.org/10.1145/1774088.1774348
58. Miche, M., Schreiber, D., Hartmann, M.: Core services for smart products. In: Proceedings of Smart Products: Building Blocks of Ambient Intelligence (Am I-Blocks'09), collocated with AmI'09 (2009)
59. Aitenbichler, E., Kangasharju, J., Mühlhäuser, M.: MundoCore: a light-weight infrastructure for pervasive computing. Pervasive Mobile Comput. **3**(4), 332–361 (2007). http://elara.tk.informatik.tu-darmstadt.de/publications/2007/aitenbichler07pmc.pdf
60. Bellifemine, F., Rimassa, G.: Developing multi-agent systems with a FIPA-compliant agent framework. Softw. Pract. Exper. **31**, 103–128 (2001). http://dx.doi.org/10.1002/1097-024X(200102)31:2<103::AID-SPE358>3.0.CO;2-O
61. Pokahr, A., Braubach, L., Lamersdorf, W.: Jadex: a BDI reasoning engine. Multiagent Systems, Artificial Societies, and Simulated Organizations, vol. 15. Springer (2005)

62. Aiello, F., Fortino, G., Gravina, R., Guerrieri, A.: A java-based agent platform for programming wireless sensor networks. Comput. J. **54**(3), 439–454 (2010)
63. Fortino, G., Guerrieri, A., O'Hare, G., Ruzzelli, A.: A flexible building management framework based on wireless sensor and actuator networks. J. Netw. Comput. Appl. **35**, 1934–1952 (2012). http://www.sciencedirect.com/science/article/pii/S1084804512001683
64. Bellifemine, F., Fortino, G., Giannantonio, R., Gravina, R., Guerrieri, A., Sgroi, M.: SPINE: a domain-specific framework for rapid prototyping of WBSN applications. Softw. Pract. Exp. **41**, 237–265, 03 (2011). http://dx.doi.org/10.1002/spe.998
65. Evans, D.: The internet of things: how the next evolution of the internet is changing everything. CISCO white paper (2011)
66. Kephart, J., Chess, D.: The vision of autonomic computing. Computer **36**(1), 41–50 (2003)
67. Pujolle, G.: An autonomic-oriented architecture for the internet of things. In: IEEE John Vincent Atanasoff 2006 International Symposium on Modern Computing, 2006 (JVA '06), pp. 163–168 (2006).
68. Cheng, Y., Farha, R., Kim, M.S., Leon-Garcia, A., Hong, J.W.-K.: A generic architecture for autonomic service and network management. Comput. Commun. **29**(18), 3691–3709 (2006). http://www.sciencedirect.com/science/article/pii/S014036640600226X
69. Xu, X., Bessis, N., Cao, J.: An autonomic agent trust model for iot systems. Procedia Comput. Sci. **21**(0), 107–113 (2013). http://www.sciencedirect.com/science/article/pii/S1877050913008090
70. Fortino, G., Parisi, D., Pirrone, V., Fatta, F.I.: Bodycloud: a saas approach for community body sensor networks. Future Gener. Comput. Syst. **35**(6), 62–79 (2014)
71. Fortino, G., Russo, W.: Towards a cloud-assisted and agent-oriented architecture for the internet of things. In: Proceedings of the 14th Workshop From Objects to Agents (WOA), vol. 1099, pp. 60–65 (2013). http://CEUR-WS.org
72. Fortino, G., Russo, W.: ELDAMeth: an agent-oriented methodology for simulation-based prototyping of distributed agent systems. Inf. Softw. Technol. **54**(6), 608–624 (2012). http://www.sciencedirect.com/science/article/pii/S0950584911001960

Mobile Agents-Based Smart Objects for the Internet of Things

Teemu Leppänen, Jukka Riekki, Meirong Liu, Erkki Harjula and Timo Ojala

Abstract We propose mobile agents for enabling interoperability and global intelligence with smart objects in the Internet of Things, with heterogeneous low-power resource-constrained devices where the systems span over disparate networks and protocols. As the Internet of Things systems are in continuous transition, requiring software adaptation and system evolution, an adaptable composition is presented for the mobile agents. The composition complies with the Representational State Transfer principles, which are then utilized in agent creation, migration and control. Moreover, the smart objects' resources, their capabilities, their information and provided services are exposed to the Web for human-machine interactions. We consider the requirements for enabling mobile agents in the Internet of Things from multiple perspectives: the smart object, the mobile agent and the system. We present interfaces for smart object internal architecture to enable mobile agents and to enable their interactions. An application programming interface is suggested with a system reference architecture, which includes components in the information infrastructure. Lastly, an evaluation metrics for the mobile agent composition and for the smart objects' resource utilization are suggested, taking the different types of system resources and their utilization into account, assisting in the system, application, smart object and the mobile agent design.

Keywords Smart object · Mobile agent · Representational State Transfer · Interoperability · Internet of things · Human-machine interaction

T. Leppänen (✉) · J. Riekki · M. Liu · E. Harjula · T. Ojala
Department of Computer Science and Engineering, University of Oulu, Oulu, Finland
e-mail: teemu.leppanen@ee.oulu.fi

G. Fortino and P. Trunfio (eds.), *Internet of Things Based on Smart Objects*, Internet of Things, DOI: 10.1007/978-3-319-00491-4_2,

1 Introduction

The Internet of Things (IoT) refers to a globally connected, highly dynamic and interactive network of physical and virtual devices [2]. The IoT requires integration and collaboration of disparate technologies in wireless and wired networks, heterogeneous device platforms and application-specific software. These IoT technologies include, but are not limited to, Ethernet, RFID, BlueTooth, Wi-Fi, ZigBee and 6LoWPAN. IoT systems require scalability beyond millions of devices, where centralized solutions could reach their bounds. To achieve global connectivity, standardized protocols and interfaces are necessary to address device heterogeneity and to enable universal resource access. Moreover, IoT systems cannot have fixed deployments or fixed system configurations, as the environment is in continuous transition. Runtime deployment of new services and applications has to be supported in IoT. As the systems evolve over time, it is necessary to consider software adaptation and evolution in order to cope with environment, system configuration and application requirement changes. Specifically, the issues here include interoperability between different standards, protocols, data formats, resource types, heterogeneous hardware and software components, database systems and human operators [2, 9]. The IoT paradigm transforms everyday physical objects into wirelessly networked, intelligent, autonomous and self-aware smart objects and enables smart objects to observe their environment through integrated sensors, store the information and interpret it proactively, cooperate by sharing information and react to changes in the environment [7, 11, 16]. The IoT then becomes a loosely-coupled and decentralized system of smart objects, based on distributed applications and global intelligence.

The agent-based systems feature decentralization and flexibility in the system configuration and allow abstracting heterogeneous subsystems and system resources for cooperation [3, 4, 7, 9, 10, 12, 16]. Agents act autonomously, possess self-properties and allow the direct manipulation of the hosting device and its physical components, such as sensors and actuators. Additionally, in agent-based systems, communication and information processing costs can be reduced by distributing information processing closer to actual data sources. Mobile agents are autonomous programs that transmit their execution state from device to device in networked systems [17], which provides means for software adaptation, system evolution, and tolerance to system or environment failures. Furthermore, mobile agents enable the dynamic reuses of hardware components and asynchronous execution of computational tasks.

This heterogeneous shared environment is not only a technical system, but IoT devices and smart objects also interact with human users. Therefore, smart user interfaces [9] are required for humans to interact with smart objects and access various services and use applications. Abstracting the system through RESTful Web services leads to the vision of Web of Things (WoT) [8], where each smart thing is equipped with a tiny Web server according to its capabilities, which then becomes an integral part of the Web. For embedded networked devices, Web connectivity can be enabled with embedded Web services [18], or by the smart gateways abstracting the most low-power resource-constrained devices, such as 8-bit microcontrollers

[12]. The Web is beneficial for human-machine interactions, as various services for information searching, aggregation and visualization are available.

We propose a method for the integration of mobile agents and smart objects in order to facilitate cooperation and global intelligence, extending our previous work on the mobile agent based integration in IoT systems and wireless sensor networks [12]. This leads to a number of research questions [7, 11]: (1) How are the distributed IoT system architecture designed and eventually deployed? (2) What are the middleware and programming models? (3) How to represent the smart objects capabilities and the distributed intelligence? (4) How to combine a coherent collective application with the distributed application logic? (5) What is the level of human involvement? We seek to answer these questions by enabling platform- and programming language-independent mobile agents in an open standards based framework and information infrastructure. Interactions with humans are facilitated by seamless integration into the Web.

In this chapter, we first present system design considerations for mobile agent-based smart objects in IoT and outline the requirements for the smart objects, agents and systems to enable cooperation. A mobile agent composition is then presented with a RESTful Web service API for smart object internal architecture and reference system architecture. Finally, an evaluation method is presented to assist in system and agent composition design, taking into consideration the different types of IoT system resources and their dynamic utilization.

2 System Design Considerations

IoT system architecture or middleware should facilitate general and non-specific design solutions for applications, because the systems are in continuous transition, where the system configurations and network topologies are ad-hoc, thus cannot be fixed in deployment. The IoT system should provide the augmentation of smart objects with internal and external services, object management capabilities and means for system adaptation and object evolution [6]. Then, the smart objects can control the flow of information, supporting global intelligence autonomously and cooperatively [6, 16]. For IoT, Internet-based information infrastructure is needed to leverage the capabilities of smart objects for provision of services to end-users.

The object-centric systems provide one solution for the heterogeneous fluctuating environments, enabling the abstraction of hardware, software, data and physicality [6]. The distinct features of smart objects include the ability to make complex intelligent decisions in information processing locally and to provide services for end users. In [11], three types of smart objects were presented: activity-, policy- and process-aware, with different levels of information processing and interaction capabilities. The individual different capabilities of the smart objects need to be exposed to the system, be discoverable and queried through well-defined interfaces. In [4], three types of multi-agent platforms for devices with limited resources was described, such as computing power, memory, limited user interface or real-time constrains.

First, the portal platforms do not execute the agents on resource-constrained devices, but only provide user interaction, sensing and actuation capabilities. Secondly, the embedded platforms execute the agents entirely in the device itself and, thirdly, the surrogate platforms execute the agent partially on a resource-constrained device and partially on the other devices. These high-level design considerations lead to some key challenges. The smart objects need to be globally identified and addressable. The resource access interfaces and object capabilities need to be globally discoverable. The dynamic availability of the capabilities in the smart objects is crucial in IoT system deployment. The application development should be decoupled from the smart object development as the smart objects take different roles and interaction models in the system, based on their capabilities and the agents they are hosting. Intelligent decision-making is required for information processing, but also in ad-hoc networking, data routing and providing inter-network relationships.

In Resource-Oriented architectures (ROA) [15], the main abstraction is a resource, referenced through its unique URI. The ROA is based on the principles of the Representational State Transfer (REST), which include separation of concerns with clients and servers, stateless communication, addressability of resources, link-based connectedness, uniform interface for resource access and various representations of the state of the resource. The individual capabilities and resources of smart objects, the state and composition of the agents, system and external services and application-specific tasks should all be considered as REST resources with URI. These resources are exposed to the system and be utilized through a uniform interface. The transportable URI identifies system resources and the URLs support resource hierarchies, linked resources and even private network overlays within the application, a particular task or an agent composition. The transportable URI enables to discover the smart objects, their relationships and contextual situations. This realizes the information infrastructure and system-, application- or task-based network structure repository [16]. Moreover, the IETF Constrained RESTful Environment Working Group [18] has published Internet drafts to enable embedded Web services in the low-power resource-constrained embedded networked devices. These drafts are crucial for IoT solutions, as the existing solutions based on HTTP and Simple Object Access Protocol (SOAP) may be too heavy for the most resource-constrained embedded devices. The CoRE framework enables direct access to the resources in embedded networked devices from the Web and facilitates limited human-machine interactions. These drafts additionally describe a number of infrastructure services such as resource directory facilitator and proxies for protocol translations, which are utilized to implement parts of the information infrastructure in this work.

In our previous work [12], we proposed a REST-based adaptable mobile agent composition, where the principles of REST are utilized in agent creation, composition, migration and control, realizing the requirement for single protocol in the agent transfer, messaging and control. The agent composition itself can be considered as a system resource, promoting re-use, and adaptable to react to unexpected system environment changes. With the RESTful Web services, we are able to utilize standardized uniform interfaces and communication primitives with heterogeneous IoT systems, smart objects and device platforms, based on loosely-coupled and flat distributed system architecture in WoT. Seamless integration to external Web services follows from the rule of extending the systems over the Web.

3 Requirements for Mobile Agents in IoT

We gather the requirements to enable the system-wide interoperability of smart objects with mobile agents and heterogeneous resource-constrained object platforms as well. General requirements for smart object middleware are previously presented in [5], however we consider the REST principles in the agent creation, migration, control and its composition. Extended from our previous work in [12], the smart objects are capable of sensing and actuation, storing information, local decision-making, interacting with each other and with external entities and finally, of operating in ad-hoc networks [16].

3.1 Requirements for Smart Objects-Enabled Platforms

Sensing and actuation	Smart objects are equipped with physical components, such as sensors and actuators. Both of these components should be identifiable and accessed as resources of the smart objects.
Information gathering	Smart objects can locally process the gathered information, providing them capability to understand their contexts and to make intelligent decisions.
Information dissemination	Smart objects in IoT support many interaction models, such as client-server, publish-subscribe, event-based communication and broadcast messages.
Networking	Smart objects are capable of participating in both intra-network and inter-network communications over disparate networks.
Mobile code	Smart objects in IoT need to support a number of distributed programming models: macroprogramming languages, MapReduce, code migration, task offloading, cyber foraging and virtual machines.
Shared resources	Smart objects maintain their resources and capabilities, which include gathered and refined information, object's capabilities and furthermore hosted agents' resources. The resources are be then cooperatively utilized by other objects and agents in their operations.

3.2 Requirements for Mobile Agents

Shared resources	The agents maintain their own state, exposed by the smart object. As the task state is not tightly-coupled into a physical device, the task state is cacheable and agents provide limited robustness in case of failures.
Agent composition	Agent implementation should be platform- and programming language-independent to address the software and hardware heterogeneity. The agent composition should be adaptive, modified by the hosting devices. The composition can also be exposed as a system resource.
Lightweight composition	Agents must be lightweight in composition, serializable and transferred as a whole or as sequential parts. Agents should be executable in platforms with limited processing power, memory, communication capabilities and battery lifetime. Binary message formats are a necessity for most of the resource-constrained embedded devices.
Dynamic deployment	The agent life-cycle is application-dependent.

3.3 Requirements for IoT Systems

Standardized interfaces	Standard, unified and simple interfaces are required to address device heterogeneity, resource abstraction, and for universal access. To simplify the implementation of mobile agents, agent transfer, messaging and control protocols should be integrated into a single protocol, based on basic communication primitives.
Abstracted objects	Smart objects and their resources should be utilized through basic and standardized communication primitives with unified interfaces, where the primitives should be interface and protocol independent. The agents can also provide primitive or atomic operations in the system.
Abstracted resources	In addition to the resources in smart objects, the system resources include internal and external services, which may register themselves into the system with various roles, such as data producer, aggregator or interaction enablers. These resources may also introduce their own restrictions and priorities.
Dynamic deployment	The systems are in continuous transition, therefore the runtime injections of objects and agents into the system are common, where the i.e life-cycle is application-dependent.

Table 1 Mobile agent composition for the smart objects

Segment	Elements	
Metadata	Name	Agent i.e. resource name or URI
	Migration	Policy identifier
	Authorization	Access rights
	Timestamp	Time of last state update
Code	Type identifier	Task code
	Reference	URL
Resource	Local	URL list
	Remote	URL list
	Static	URL list
	Reference	URL list
State	State variable list	
	Local variable list	
Historical data	Variable list	
	URL list	

Dynamic binding — The objects are simultaneously acting as servers for their local resources and as clients for the resources in other objects. The agents should allow dynamic binding to resources and dynamic mapping of the task into any system configuration. An agent composition should, in general, be exposed to the system by the devices and be adaptable. Runtime lookups and loose coupling to the resources are facilitated by stateless communication.

Scalable configuration — Scalability beyond current networked systems is required. Thus distributed architectures with loosely-coupled services become necessity. Gateways and proxies are introduced to abstract heterogeneous subsystems, spanning over networks, protocols and communication interfaces.

4 Composition for the Mobile Agents

We extend the agent composition presented in [12], to fulfill the smart objects and system requirements. The composition, illustrated in Table 1, consists of three segments: code, resource and state. In addition the composition includes metadata, such as a unique name or URI, to register the agent into the system and to enable resource lookups. The metadata also contains a description of globally known migration policy to control the agent migration. The metadata may also include the last time the agent state was updated and authorization information for the agent and the required resources. With proper authorization, we allow smart objects, smart gateways and proxies to modify the composition to adapt and evolve to system environment, application or task configuration changes and to dynamic resource availability. Then, the agent composition itself becomes a system resource. Historical data can be included in the composition for agent tracking purposes.

4.1 Code Segment

The agent task code is stored into this segment. The code can be presented in any programming language: high-level macroprogramming language, scripting language, precompiled binaries, bytecode or even as machine language instructions. The segment allows multiple code segments for multiple heterogeneous platforms, which requires an identifier of the code type. Additionally, to minimize the composition size, the segment can contain a reference to the code in the system repository for on-demand code retrieval.

4.2 Resource Segment

The resource segment lists the local, remote and static resources for the task execution. The local resources refer to the resources exposed by the hosting smart object, whereas remote and static resources are external to the object. The remote resources are accessed each time the agent migrates or the task execution iterates. The static resources are remote and constant for the lifetime of the agent, hence the representation is requested only once and moved into the state segment as a variable. How the object binds to the remote resources are determined by the references and the resource access interfaces. If the resource segment is a reference, resources are requested from a application-specific or global system service. This allows sharing the segment becomes a system resource and enabling runtime modifications as well. The resource segment therefore presents dynamic and partial view of the system resources utilized by the agent, as an overlay.

4.3 State Segment

The state segment contains the current state of the agent, i.e. the intermediate or final results of the task. The state is then returned as the agent resource representation. Other local data, such as a program counter, local variables and retrieved static resources are stored in this segment as well [12].

4.4 Historical Data

This segment is optional. It contains the previous states of the agent, its local variables and previously visited locations, for tracking the agent and its behavior. Also, to minimize the composition size, this segment may contain URLs to a repository hosting this information.

4.5 Agent Mobility

The local resource segment dictates where the agent migrates in the system, with the particular migration policy given in the metadata segment. We can utilize any migration police, for example: (1) the agent visits the objects listed in the local resource segment each only once, (2) the agent considers the local resource segment as a ring buffer circulating though the devices, (3) the agent message is broadcasted to all the objects at once, (4) the local resource segment lists gateways or proxies, which distribute the agent to any number of abstracted smart objects. Actually, the migration policy can also be considered as a system resource. Whenever the local resource segment is a reference, the smart objects rely on the information infrastructure, such as the network structure repository [16] for migration instructions.

This agent migration procedure requires, at minimum, that the agent is cloned in the host device and sent to the new host, where the state is first updated by executing the computational task. Then the host registers the agent into the system name server or directory facilitator. Here the updated agent state is exposed into the system and other objects can access the new state in the new host. After successful registration, an acknowledgement is sent to the previous host, which can then delete the agent from its memory and free the utilized resources.

4.6 Implementation Considerations

Considering different utilization of the adaptable agent composition, we assume the following. The state segment cannot be omitted, as it is the agents resource representation. If the code segment is omitted, then the agent works as a data aggregator, migrating through the listed smart objects. Moreover, this enables event-based communication [1] as the agent composition can be considered an event with a state. If the local resource segment is omitted, the agent does not migrate autonomously, which implements task offloading. The remote and static segments are optional based on the required resources. All the activity-, policy- and process-aware smart object functionality [11] become now possible. This agent composition inherently supports the multi-agent platform types in [4], as the requested resource representations dictate which parts of the agent task code are executed and where.

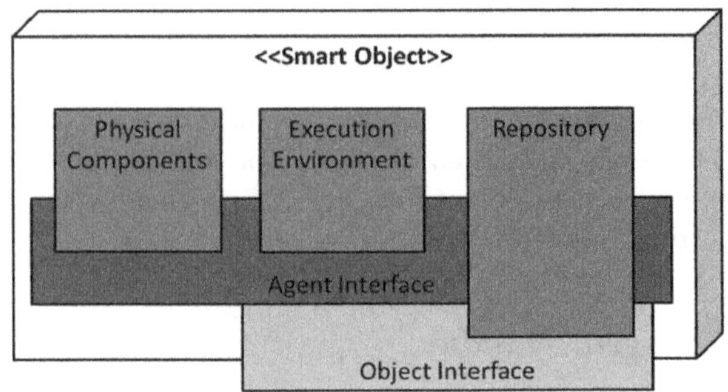

Fig. 1 Smart object internal architecture to facilitate mobile agents

With the flexible structure of the resource segment, objects and intermediates can modify the composition [12]. The client-server paradigm is the default; agents send state and resource requests to other agents. Publish-subscribe paradigm can be achieved through mobile agents as events. MapReduce can be implemented by cloning the agent, or by broadcasting the resource requests to the system devices, where partitioning the task into smaller computational units can be considered before sending the task to the devices. Macroprogramming languages can be supported through agents as high-level code abstractions or as code primitives, which can be introduced to the system as on-demand task code or as global system resources. These primitives representations can be considered remote methods or in-line code for the task.

Mapping the agent composition to different protocol messages needs to be considered. With the HTTP, we can assume the composition for example as HTML or XML document, EXI XML representation or as JSON object. However, these human-readable formats may introduce too much overhead in communication with low-power resource-constrained embedded devices. Therefore, we presented the agent composition mapping into a significantly smaller, in size, binary Constrained Application Protocol (CoAP) message structure in [12].

5 Smart Object Reference Architecture

For the smart objects, we identify three software components, which are necessity to enable mobile agents: the execution environment to run the actual agent task, a repository to store the resources in these objects and the physical components, such as sensors and actuators. The repository contains both data and the knowledge base, typically in a relational database. We also define two interfaces: the agent interface to enable the handling of mobile agents and the object interface for communication with other smart objects and the system. See Fig. 1 for the proposed smart object internal architecture.

The execution environment (EE), it is a hardware and operating system dependent. The EE is capable of querying information from the repository and from the physical components to compile a runnable code for execution, after retrieving the required resource descriptions. Additionally, EE provides methods for actuating and controlling the physical components. The EE must feature a method to immediately stop the agent code execution, called by the EE or by the agent task code, which enables the agent to control its own execution.

The implementation of EEs for the Android operating system in Java and for Atmel microcontrollers in the C programming language was presented in our previous work [12]. The Android EE allows scripting languages Python and JavaScript as agent task code, where a language-specific engine is invoked to execute the script code. A HTTP server component is used for communication and a SQLite database for the repository. The EE in the microcontrollers uses IntelHEX precompiled code for agent task code. The task code is flashed into the memory in the device, as code cannot be run from the RAM in the ATmega architecture. However, the architecture allows flashing program memory sections without a reset, a crucial feature here. The local and remote resource representations are stored into a shared memory block in RAM, from where the executable code accesses them as 16-bit variables through common pointers. These pointers and the API methods are defined in a common C header file. In the program memory, a number of slots are reserved to store the agent code and it is accessible until overwritten. The communication API was implemented in C for the CoAP protocol.

5.1 The Agent Interface

The interface is internal within the smart object, providing the methods for handling the agent messages and agent composition, the execution of the agent tasks and local resource queries from the repository. Methods are provided to control the integrated physical components and to stop the agent execution immediately.

Marshal/Unmarshal	Handles the serialization and deserialization of the agent composition into an internal data structure in the device memory and back into the transferable agent composition, then utilized by the object interface. The data structure stores the binding of the remote and local resources.
Map/Unmap	Maps the internal data structure and local resource representations into the executable code object. After the task execution, the internal data structure is updated with the new resource representations.
Execute	Runs the executable task code object.
Getter	Retrieves the intermediate state of the agents task from the internal data structure, to respond to external state queries.

Poster — Used for disseminating events from the task code and for actuating the physical components from the code.

Stop — Called by the EE or from the agent task code to stop the task execution and to immediately transfer the agent.

5.2 The Object Interface

This interface provides functionality for inter-object communication, including resource access and registration into the system. The methods allow query parameters for retrieving information with location information, different granularity and historical data from the repositories in the objects. This allows discovering nearby smart objects and resources dynamically.

Post — Transmit the agent between smart objects, according to the resource segment addresses.

Get — Enables two-way communication by responding to the external queries of local resources, including the agent state. Secondly, it is used to request remote and static resource representations from other objects. It may be needed to first perform resource lookup into the name server or directory facilitator.

Delete — Deletes the resource, including the agent, from the hosting object or from the system.

Register/Unregister — Registers the object, its resources and capabilities into the system. Unregister is used to remove the resource description from the system. Whenever an object is hosting an agent, its identifier with the object network address is registered into the system. The address of the name server or directory facilitator should be globally known by all system components.

6 System Reference Architecture

The system reference architecture is generally based on the framework described by IETF CoRE Working Group in [18]. The benefit of the CoRE framework is that it allows embedded Web services, i.e. Web connectivity, for the most resource-constrained embedded networked devices.

In the Fig. 2, the resource directory (RD) acts as a name server and stores the resource descriptions as a part of the information infrastructure. Smart objects can lookup exposed resources in the system from the RD by the presented API. As described in [13], the RD can be a part of an P2P overlay over the IoT system. Queries can be based on URI or resource name, output type, semantic interpretation, and both virtual and physical location. Secondly, in the system architecture we

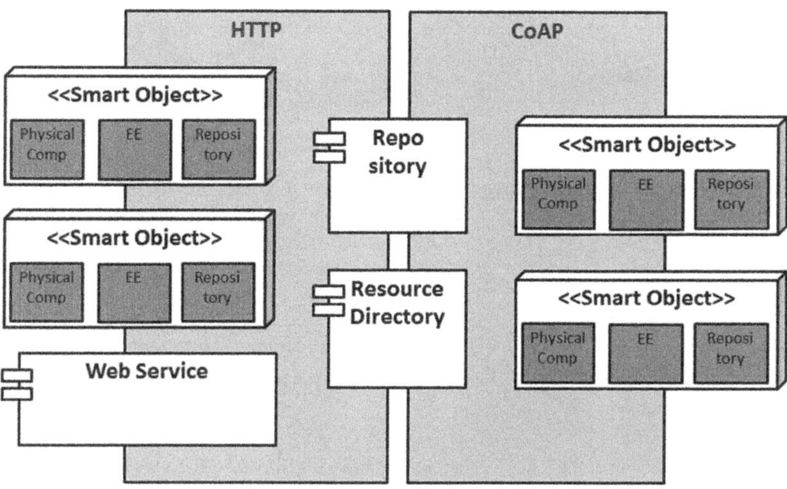

Fig. 2 Smart object-based IoT system reference architecture

utilize generic Web service in several different roles. Web service can abstract heterogeneous hardware and software technologies, heterogeneous systems and smart objects resources, coordinate application or task execution to provide application-specific intelligence, expose external services into the system, work as a gateway or proxy over disparate networks enabling interoperability and finally facilitate human-machine interactions over the Internet. Lastly, we utilize a repository component to store and expose global and application-specific resources into the system, such as agent task codes and agent compositions, accessed through the RD lookups. Therefore, the repository enables smart objects to adopt any agent-based role and facilitates the re-use of system resources as a part of the information infrastructure. The repository can also be part of, for example, a P2P overlay. These system components could provide communication interfaces for different protocols, namely HTTP and CoAP, to allow access over disparate networks.

7 Application Programming Interface

We extend the API presented in [12] with smart objects-based features for a reference RESTful API for mobile agent-based application development, complying with the ROA. With this method, the role of the objects depends solely on their local resources and on the agents they host. The API features mobile agent creation and control, agent migration, communication between devices and agents, and also local and remote access to the resources in the smart objects. It facilitates interobject and interagent

Table 2 Application programming interface

GET/object/resource GET/web_service/resource	This method is used for requesting resource representations either from a smart object or from a Web service
POST/object/agent_name POST/web_service/resource/agent_name	With the first method, the agent migrates between objects. In the second case, an agent is injected into the system for the given resource, exposed by the Web service. If an agent composition is not provided in the message body, a resource lookup is done to locate the smart objects hosting the particular resources. The resource segment is compiled automatically, if resources based on the agent name are available in the system
DELETE/object/resource DELETE/web_service/resource	This method will delete the given resource from the addressed smart object and request the deletion of the object from the RD. In the second case, a lookup to locate the hosting object for the resource is first performed, if needed to locate the hosting agent or service
GET/repository/agent_name POST/repository/agent_name	These methods will retrieve from or store the particular agent task code into the repository, with the given platform identifier as a query parameter. When adding new code, included in the message body, the repository will register the task as a system resource
GET/resource_directory/resource POST/resource_directory/resource	The lookups to the resource directory follow the same methods: GET returns the description of a resource and POST injects new resource description from the message body to the directory. The resource description follows the format outlined in [13]

Table 3 Additional query parameters

object={list of smart object URIs}	Allows directly manipulate the resources, including agents, in these particular smart objects. The requests are only sent to these particular devices. This can also override the resource segment in the agent composition
location={URI}	This identifier can be both physical location or logical address of the resource
time={start_time, end_time}	Access historical information in the agent, smart object or system resource. Additionally, when this parameter is sent to the RD with lookup request, it allows tracing the particular object or resource
rate={integer}	Set the granularity in the information requests, if available. This is an application-specific parameter

communications with basic HTTP and CoAP methods, additionally with inherent content negotiation and authorization methods. This realizes the requirement for a single protocol with a uniform interface. See Table 2 for the API description. We introduce additional parameters, in Table 3, for querying historical data with different information granularity and tracing the agent or objects location or status.

8 Evaluation of Agent-Based IoT Systems

To evaluate mobile agent-based smart objects in IoT, with the possibility of dynamic resource utilization, a set of specific measures are needed. We propose here metrics such as resource utilization costs in terms of access and communication latencies, to describe system-, application-, device-, object- and agent-based characteristics. In our earlier work [12], we proposed communication, remote resource access and agent migration latencies and computational overhead in agent task execution as the measures. Furthermore, we can compare different configurations of the above.

1. With the resource access latencies, we measure the latencies either directly between heterogeneous devices or through the abstracting Web services. Additionally, we should measure the access latency from the Internet to system platforms, considering resource access over disparate networks. This measure could include the request processing time in the hosting device. In the agent composition, we assume that this latency is dominated by the number of remote resource queries and should linearly increase as in [14]. Queries to the RD are considered the same as standard resource accesses.
2. With the computational overhead, we measure the computational latency in the particular EE in executing agent task code. Platform-specific latencies include time for system atomic service invocation, marshaling and mapping the composition into the device memory, running the code and composing the agent message again.
3. Agent migration latency includes the overhead of agent registration into the RD by the hosting device, sending the agent as a message to the next device and waiting for acknowledgement, after which the agent can be deleted from the memory. This does not include the computational overhead or additional resource access latencies. This latency would increase with introducing security measures, such as guaranteed reliable message transmission, and with large-size agent composition.

However, we found out that in the real-world environment [12, 14], conclusive evaluation would be difficult to conduct with heterogeneous smart object platforms, as the system configuration, device and object deployment, agent composition, required resources and their locations are largely application-specific. Additionally, the varying communication latencies, changing network conditions, device failures and resource availability are difficult to consider. The evaluation additionally introduces overhead, which would reduce query response times in the most resource-constrained platforms, such as wireless sensor networks nodes. Therefore, we should utilize of indirect evaluation, for example in smart object or agent communication through application-specific Web services or by tracing the resource accesses or agent migration indirectly from the RD. Echo request message's, such as the ping, round-trip latencies could be useful for measuring communication latencies as a baseline.

To assist in the evaluation of mobile agent-based IoT system, application and agent composition design, we proposed simplified equations in our previous work [12]. We extend the equations to include smart object-specific features, such as the repository component in each device. The equations identify the relevant factors in each case, and with modifications allow calculating application-specific costs with different system and resource configurations. In Eq. 1, we estimate the maximum latencies C in particular execution environment k, including the resource access, executing the agent task and the following agent migration latency. Here r is the number of remote resources, T_r is the response time for remote and static resource requests, T_k is the computational overhead in the device and T_m is the migration latency from sending the agent message to receiving an acknowledgment message. T_r is added once for agent registration to the system. The local resource query latencies T_l can be considered negligible, however with large information chunk retrieval or with large number of local resource accesses l, this can be considerable. Based on the observations in [12, 14], the remote resource queries and migration latencies dominate these costs.

$$C_k = (r+1)T_r + lT_l + T_k + T_m \tag{1}$$

The Eq. 2 gives the total agent migration costs C_{Total}, for a particular mobile agent-based service as the agents migrate over disparate networks.The additional latency for static resource queries is included, where s is the number of static resources. The number of disparate networks is d, and here it is assumed that the agent migrates only once to each network. The equation can be modified to cover different scenarios. The agent migration time between networks is given as $T_{m,d}$. We include the latency of possible message translations in the gateways as T_p. The latencies in each execution environment are given in $C_{n,t}$, from Eq. 1, where the number of devices running each execution environment is n.

$$C_{Total} = sT_r + \sum_{d-1}(T_p + T_{m,d}) + \sum_{d}\sum_{n-1}(C_{n,t}) \tag{2}$$

The cost, as latencies, is dominated by the number of platforms in each network and the previously noted remote resource accesses [12, 14]. Therefore, the remote resources in the system design and agent composition should be considered as static or local resources as much as possible [12]. However, this is an application-specific tradeoff between the agent migration costs and resource access latencies, as the composition allow the different utilization of the resources. In IoT, we can envision systems over a number of disparate networks, all with their own characteristics and technologies, therefore, the migration cost and resource access latencies over disparate networks are significant factors in the system and application design and in the deployment phase.

9 Related Work

To start with, an extensive evaluation of middleware for smart objects and smart environments in IoT, can be found in [5]. Here we consider the previous work related to agent-oriented smart object-based systems in IoT.

In [9], the authors envision agent-based IoT system architecture, where the resources are represented by agents. Agents handle monitoring the state of the resource, historical data storage and the interactions with other components and humans. Monitoring and coordination of the resources is done through specific roles played by the agents. Communication is based on the role of the agent and not to its name or identifier. For resource discovery, the semantic queries are addressed to the directory to locate the resource identifier. The tasks are written in a rule-based language, where the agents provide the system configuration for the tasks and react to configuration changes.

In [7], the authors present agent-based architecture for smart objects, where the IoT system heterogeneity is abstracted with layers. The system architecture provides communication middleware to abstract underlying details, a component for managing communication with external systems, a resource discovery module, adapters to abstract sensors and actuators as system resources, and lastly components for managing contexts, knowledge base and reasoning. The implementation is Java-based. The master-slave model is used with smart objects, where a coordinator manages the set of software entities, running on other smart objects. The coordinator controls the hosting smart object though internal communication protocol and is the sole component to communicate with other smart objects or system devices, through external communication protocol. An internal software framework provides API for atomic services and runs the EE as an additional internal software framework.

In [16], an event-driven smart object framework for IoT is presented. The smart objects communicate by forming ad-hoc clusters, based on the common context of objects, with electing representative to each cluster. The objects communicate within the cluster, where only the representative communicates with the infrastructure. In communication, XML documents are disseminated over SOAP and HTTP through a Web service in a gateway node. Two types of events have been defined: network structure changes to manage the clusters and events to disseminate sensor data. For addressing and routing, the authors have developed their own mechanism, considering merge and split operations with unique and reusable addresses. The role of cluster representative rotates according to available resources in objects, balancing the communication load.

In [1], an agent-oriented and event-based framework for cooperative smart objects, based on the architecture in [7], is presented. Smart objects' behaviour, in the form of tasks, is separated from the event-based communication management. The tasks are separated as system tasks, providing basic services for the smart objects and as user-defined tasks, in application-level, to define the smart objects' behaviour as plans. Events are categorized as information, request, log and error. Event types include system internal events, external events and another smart object as an event source. The communication model is publish-subscribe, where each smart object publishes its topics and services for others to utilize.

In [3], interoperable agents in IoT are presented, abstracting heterogeneous devices and communicating over different access technologies simultaneously. The agents register their identifier, type and transport protocols to the directory facilitator. The facilitator enables the registration and discovery of agents, group memberships of agents, system services and a messaging service for messages between agents. The groups, as IoT applications, enable multicast messaging with the members.

Considering agent platforms for smart objects, the authors in [10] present a multi-agent platform for embedded systems, based on the Java virtual machine. The device platforms include static system agents providing interfaces to the system services and, on the other hand, dynamic service agents running the smart home applications. In [1], mobile agent framework for SunSPOT platforms is implemented in Java. The agents are modeled as multi-plane event-based state machines, where the state transitions come in response to events. New events can then be emitted asynchronously. A distinct feature is the timing of the agent operations by system components, which additionally offer services for communication and agent control.

In comparison, we presented a novel, language- and platform independent composition for mobile agents-based smart objects. This method is based on open standards for communication over disparate networks and for collaboration support without specific interaction models or middleware. The information infrastructure is realized with the IETF CoRE framework [18], additionally enabling resource-constrained device platforms for smart objects becoming integral part of the Web. The system architecture is flat and is not restricted to specific interaction, communication or programming models. Centralized system configuration or agent coordination is not facilitated and we do not apply any specific system configuration or task plan with the smart objects. Instead, we expose the modifiable agent composition into the system as a common resource. This method facilitates dynamic interlinked many-to-many communications, including external systems, despite the roles of the agent or smart objects. The REST design principles and unified interfaces are utilized for agent creation, migration, messaging, control and exposing system resources to the Web. Lastly, although Java software components are modular, portable and provide object-oriented features for programming, Java virtual machine-based solutions may be too heavy for the most resource-constrained embedded devices.

10 Discussion

In this work, we proposed a method for integration of autonomous smart objects with mobile agents, with open standards for communication and cooperation support without a specific middleware solution. We presented language- and platform-independent mobile agent composition, which enables global intelligence and different interaction models for the smart objects and mobile agents. The roles of the smart objects are decided by the agent composition, which promotes the dynamic re-use of the system resources with different simultaneous applications. Mobile code is inherently supported as the mobile agents can be considered as application-level tasks, high-level programming abstractions and code primitives. The expected benefits include: mobile agents enable global intelligence, mobile agents facilitate adaptable system configurations and dynamic service composition, distribute computational load in applications, exploit locality in communication, and finally provide re-usability and robustness for the smart objects.

The REST principles are utilized in agent creation, migration, control and, in larger-scale in smart object communication, system resource access and exposing the resources to the internet, including the agent composition itself. This realizes the single protocol for uniform interface in ROA-based architecture. Moreover, the system resources, services and smart objects are exposed to the Web for human-machine interactions, which provides integration into the WoT.

The presented evaluation method, albeit generic and simplified, can assist in application-specific IoT system design, in smart object- or mobile agent-based dynamic service composition and in system service response latency estimations. Additional system and network-specific parameters should be introduced to real-world evaluations. The inevitable security and privacy issues in agent-based approaches were omitted in this work, but to some extent the security mechanisms of communication protocols are available with RESTful Web services.

Acknowledgments This research was conducted with the MAMMotH Project, funded by the Finnish Funding Agency for Technology and Innovation (Tekes), at the Department of Computer Science and Engineering, University of Oulu, Finland.

References

1. Aiello, F., Fortino, G., Gravina, R., Guerrieri, A.: A java-based agent platform for programming wireless sensor networks. Comput. J. **54**(3), 439–454 (2011)
2. Atzori, L., Iera, A., Morabito, G.: The internet of things: a survey. Comput. Netw. **54**(15), 2787–2805 (2010)
3. Ayala, I., Amor, M., Fuentes, L.: An agent platform for self-configuring agents in the internet of things. In: Infrastructures and Tools for Multiagent Systems, p. 65 (2012)
4. Carabelea, C., Boissier, O.: Multi-agent platforms on smart devices: dream or reality. In: Proceedings of the Smart Objects Conference, pp. 126–129. Grenoble, France, (2003)
5. Fortino, G., Antonio, G., Russo, W., Savaglio, C.: Middlewares for smart objects and smart environments: overview and comparison. In: Fortino, G., Trunfio, P. (eds.) Internet of Things based on Smart Objects: Technology, Middleware and Applications, Internet of Things. Springer, Berlin (2014)
6. Fortino, G., Guerrieri, A., Lacopo, M., Lucia, M., Russo, W.: An agent-based middleware for cooperating smart objects. In: Corchado, J., Bajo, J., Kozlak, J., Pawlewski, P., Molina, J., Julian, V., Silveira Ricardo, A., Unland, R., Giroux, S. (eds.) Highlights on Practical Applications of

Agents and Multi-Agent Systems, Communications in Computer and Information Science, vol. 365, pp. 387–398. Springer, Berlin (2013)
7. Fortino, G., Guerrieri, A., Russo, W.: Agent-oriented smart objects development. In: 16th IEEE International Conference on Computer Supported Cooperative Work in Design, pp. 907–912 (2012)
8. Guinard, D., Trifa, V., Wilde, E.: A resource oriented architecture for the web of things. In: Internet of Things 2010 Conference, pp. 1–8 (2010)
9. Katasonov, A., Kaykova, O., Khriyenko, O., Nikitin, S., Terziyan, V.Y.: Smart semantic middleware for the internet of things. In: 5th International Conference on Informatics in Control, Automation and Robotics, Intelligent Control, Systems and Optimization, pp. 169–178. Funchal, Portugal (2008)
10. Kazanavicius, E., Kazanavicius, V., Ostaseviciute, L.: Agent-based framework for embedded systems development in smart environments. In: Proceedings of International Conference on Information Technologies. Kaunas, Lithuania (2009)
11. Kortuem, G., Kawsar, F., Fitton, D., Sundramoorthy, V.: Smart objects as building blocks for the internet of things. Internet Comput. **14**(1), 44–51 (2010)
12. Leppänen, T., Liu, M., Harjula, E., Ramalingam, A., Ylioja, J., Närhi, P., Riekki, J., Ojala, T.: Mobile agents for integration of internet of things and wireless sensor networks. In: IEEE International Conference on Systems, Man, and Cybernetics, pp. 14–21 (2013)
13. Liu, M., Leppänen, T., Harjula, E., Zhonghong, O., Ramalingam, A., Ylianttila, M., Ojala, T.: Distributed resource directory architecture in machine-to-machine communications. In: IEEE 9th International Conference on Wireless and Mobile Computing, Networking and Communications, pp. 319–324 (2013)
14. Malek, S., Medvidovic, N., Mikic-Rakic, M.: An extensible framework for improving a distributed software system's deployment architecture. IEEE T Softw. Eng. **38**(1), 73–100 (2012)
15. Richardson, L., Ruby, S.: RESTful web services. O'Reilly (2008)
16. Sanchez Lopez, T., Ranasinghe, D., Harrison, M., McFarlane, D.: Adding sense to the internet of things. Pers Ubiquit Comput. **16**(3), 291–308 (2012)
17. Satoh, I.: Mobile agents. In: Nakashima, H., Aghajan, H., Augusto, J.C. (eds.) Handbook of Ambient Intelligence and Smart Environments, pp. 771–791. Springer, Berlin (2010)
18. Shelby, Z.: Embedded web services. IEEE Wirel. Commun. Mag. **17**(6), 52–57 (2010)

Service-Oriented Middleware for the Cooperation of Smart Objects and Web Services

Andrea Giordano and Giandomenico Spezzano

Abstract Many physical devices can be interconnected and cooperate by Internet of Things (IoT), providing and consuming information available on the network. These will not only provide information by monitoring the real-world, but create complex collaborations, interacting also with business processes, in order to provide sophisticated value-added services. In addition, business processes can also adapt their behavior in response to real-time context updates. Web services technology offers a promising approach to provide information and functionalities of physical objects to business processes, since it facilitates interoperability and encapsulates the heterogeneity and specificity of physical objects. To address the dynamic composition of web services in a decentralized, distributed manner, with no single point of failure, a choreography execution model can be used. This chapter describes an approach to support adaptable business processes (workflows) considering changes in the state of Things; likewise, whenever needed, the software controlling the behavior of sensors can be dynamically configured as a result of changes in the functional specifications of business processes.

1 Introduction

From an enterprise and economic perspective, the Future Internet will be the basis for *a web-based service economy* [1] that will merge the digital and physical worlds. The Future Internet, it is now widely accepted, will have four pillars [2]. Besides the Internet of Networks, there will be an Internet of Services as well as Internet of Things integrating common objects into our lives. Finally, an Internet of Contents &

A. Giordano · G. Spezzano (✉)
CNR-ICAR, Via P. Bucci 41C, 87036 Rende, CS, Italy
e-mail: giordano@icar.cnr.it

G. Spezzano
e-mail: spezzano@icar.cnr.it

G. Fortino and P. Trunfio (eds.), *Internet of Things Based on Smart Objects*,
Internet of Things, DOI: 10.1007/978-3-319-00491-4_3,

Knowledge, and an Internet of People are foreseen too. It is important to note that these terms should not be regarded as different "Internets" that will exist in parallel, but rather as different aspects of a common Future Internet.

The innovative and rapidly evolving area of Internet of Things [3] and Services (IoTS) integrates two of the four pillars of the Future Internet and investigates a world where smart objects (SOs) [4–7]—that is, autonomous physical/digital objects augmented with sensing, processing, and network capabilities, are seamlessly integrated into the information system where they can become active participants in business processes. The supporting service-oriented middleware [8] will then abstract the functionalities of SOs as services as well as provide the needed interoperability and flexibility, through a loose coupling of components and composition of services.

Efforts in this area are focused on a development of platforms and solutions where services and SOs cooperate and can be employed in real-world applications in industrial domains such as manufacturing, e-Health, smart cities, home automation, e-Business, etc. However, owing to the heterogeneity of devices and tight coupling of individual information systems, developers cannot easily create their specific applications by combining physical devices and web resources. To address these problems, we proposed to realize composite applications combining services and SOs by event-driven choreographic workflows.

Nowadays enterprise systems are built on a service-oriented architecture (SOA), and business processes in such systems are modeled as an orchestration of underlying services. In order to integrate the IoT into business process systems it is necessary to service-enable IoT resources, e.g., the sensors and actuators that are used to interact with the physical environments. The current state-of-the-art is mostly focused on integration of IP enabled networked smart objects where nodes communicate their information using RESTful Web services. We argue that the approach for the integration of RESTful SOs with existing, widely deployed SOA technologies such as Web services and Business Process Execution Language (BPEL) is the key to the success of SOs in enterprise systems.

In SOA, service composition is normally achieved either through a centralized controller or in a decentralized manner. Support of decentralized workflow execution and scalability are important issues for workflow management systems since it makes it easier to obtain a flexible and adaptive composition of services.

Typically, to reach the required level of scalability, the workflow management system must be distributed and make use of replicated web services that are selected and used at runtime. Traditional workflow systems use centralized orchestration techniques which limit the scalability in the presence of a high number of services. In the orchestration model, all data pass through a centralized engine, which results in unnecessary data transfer and wasted bandwidth so that the engine becomes a bottleneck to the execution of a workflow. Choreography techniques [9], although more complex to model, offer a decentralized alternative and are optimal architectures for data-centric workflows. In this model, data are passed directly to where they are required, at the next service in the workflow.

Self-organizing in this context describes the adaptability of the model during deployment. Changes in the environment (e.g. location change, connectivity outage, reconfiguration of business processes etc.) require reorganization of the deployed components during run time to meet given Quality of Service (QoS) constraints.

Our idea is to integrate enterprise web services with RESTful SOs by exploiting the concept of service *choreography* undertaking the scalability and dynamicity issues of IoT in order to extend the existing (adaptable) service composition mechanisms. We show that applications involving SO interaction can be seen as a particular case of event-driven composite services.

To this end, the rest of chapter is structured as follows: in Sect. 2, we present our view of system architecture for the execution of event-driven workflows; Sect. 3 presents a description of the adaptive P2P agent-based framework, called Sunflower, we studied and designed, that supports autonomic management of workflows; Sect. 4 describes the integration in Sunflower of RESTful SO; Sect. 5 illustrates the details of the proposed methodology through a case study; finally Sect. 6 concludes the chapter.

2 System Architecture

This section presents an architecture for the execution of event-driven workflows (i.e. composite applications combining services and SOs through events). The inner part of Fig. 1 shows the architecture of the Sunflower service execution platform designed to support the composition of SOAP services. This cooperating model is created by WS-BPEL workflows. Sunflower permits a decentralized and optimized execution of WS-BPEL workflows upon a P2P system as described in Sect. 3.

In this context, addressing Smart Object (SO) technology and, in general, Internet of Things philosophy requires some additional mechanisms to suitably cope with physical entities. Firstly, a kind of transport layer should be chosen and implemented in order to foster proper interaction between the WS-BPEL world and concrete "things" (i.e. SO). Secondly, given that SOs capture the state of the environment in which they are embedded, environmental state modifications should be carefully handled and reflected at the Sunflower side. Finally, a mechanism is required that permits Sunflower workflows to trigger actuation upon SOs.

On the basis of the previous considerations, we propose the use of Web Service Proxy (SP) acting as an adapter/wrapper of the SO's world. Through SP, each command coming from Sunflower will be forwarded toward the SOs. In addition, each environmental state modification will be considered as an event and notified to the Sunflower part. In summary, the proposed system can be seen as a Web Service orchestrator enhanced by a SOs mashup and an even-driven engine.

SP adds the following features to Sunflower:

- support for a combination of services implemented by means of different technologies (e.g. SOAP, REST etc.);

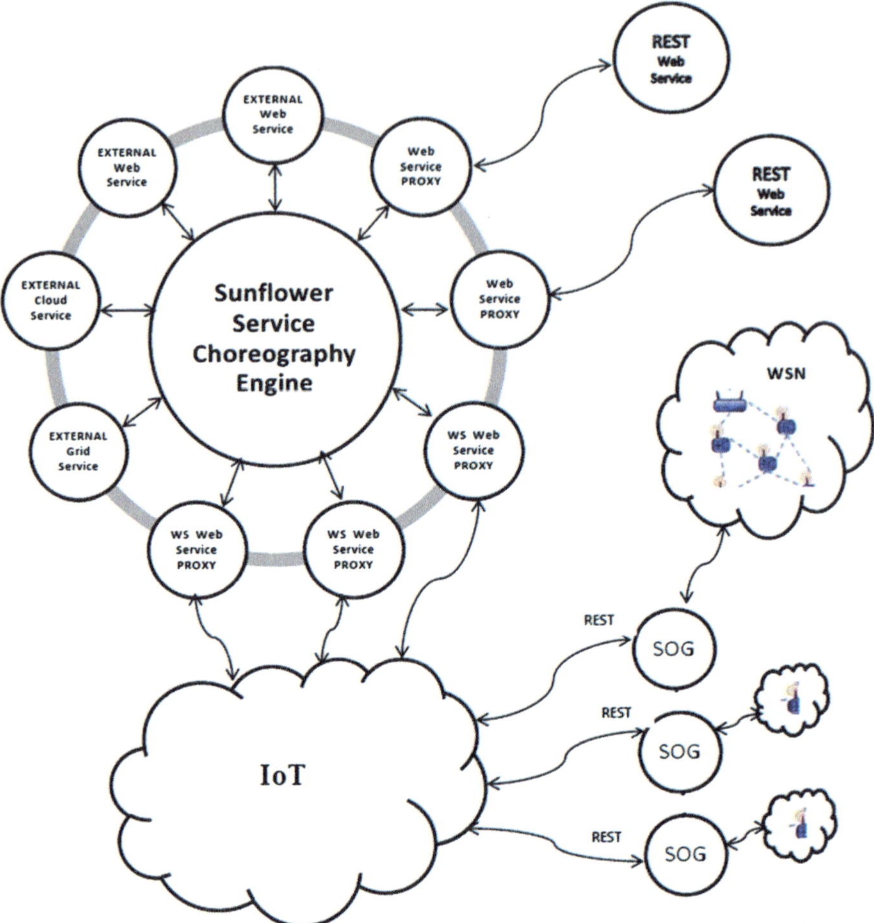

Fig. 1 Architecture for event-driven choreographic workflows execution combining Web services and RESTful Smart Objects

- interaction between Web service and SOs through an "event/action" paradigm, according to which, the occurrence of an event triggers the execution of one or more actions in other components.

SP is attached to SOs via REST invocations on a middleware layer that is in charge to manage underlying SOs. This middleware layer is represented by a so-called Smart Object Gateway (SOG). SOG offers a transparent and ubiquitous access to the physical part due to a well-established interface exposed as a REST service as described in Sect. 4.

SOG allows enterprise application to connect directly to devices without using proprietary drivers or addressing some kind of fine-grained technological issues.

In addition, it fosters the reusing of a pre-existent Web Service in conjunction with SOs thus achieving a perfect match between the Internet of Service and Internet of Things.

The low level of our architecture concerns formalization of SO and how it is integrated in the system. We define a SO as a system made up of one or more physical devices that together achieve complex behaviors. Each SO comprises "functionalities" directly provided by the physical part.

Essentially, a SO exposes an abstract representation (i.e. *machine-readable descriptions*) of the features and capabilities of physical objects spread in the smart environment. It is implemented as computer software that is used to link physical objects with the virtual world.

Functionalities exposed by different types of SOs can be combined in a more sophisticated way on the basis of event-driven rules which affect high-level applications and end-users. A SO is self-managed and self-configurable, capable of being used also out of the context for which it was initially created.

Each physical object, contained in a SO, should automatically perform a *simple* action (e.g., lighting, recording) in response to a *simple* event (e.g. detecting a user, people who sit in a chair). On the other hand, SOs must have the flexibility to change their behavior dynamically on the basis of complex applications even though they posses low processing power and small memory.

A SO changes the behaviour of physical objects by remote and dynamic reprogramming thus considering them as execution parts of business processes.

In summary, our architecture is structured in three layers:

- The first level is the SO level dedicated to the characterization of the SO abstraction in terms of its different *functionalities* that can be either sensing or actuating and can be refined by further parameters that dynamically configure it (see Sect. 4.1).
- The second level is dedicated to SOG abstraction which permits operations of remote and transparent reading and writing on SOs and, also, definition of complex rules on SO functionalities.
- The top level concerns applications written as WS-BPEL/Sunflower workflows. This level encapsulates SOs through Web Service Proxies linked via REST to the underlying layer in order to hide the heterogeneity of devices and provide a seamless way to integrate SOs with web applications.

3 Sunflower Framework

Sunflower is an adaptive P2P agent-based framework for configuring, enacting, managing and adapting autonomic workflows [10, 11]. Sunflower assumes that multiple copies of a Web service co-exist, with different performance profiles and distributed in different locations. During the execution of the workflow, when a service fails or becomes overloaded, a self-reconfiguring mechanism based on a *binding adaptation* model is used to ensure that the running workflow is not interrupted but its

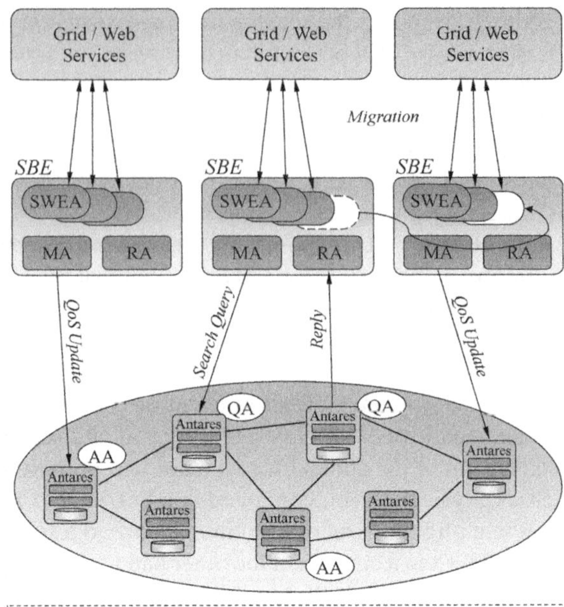

Fig. 2 The Sunflower architecture

structure is adapted in response to both internal or external changes. Figure 2 shows the architecture of the framework Sunflower.

Workflows are described in Sunflower by the BPEL language [12] in order to exploit existing design tools. Sunflower replaces the standard BPEL engine with a new decentralized engine able to exploit the dynamic information available in the network and respond to the dynamic nature of Internet.

The workflow process is enacted by a set of cooperating Sunflower BPEL engines (SBE), instantiated at all participating nodes, which are responsible for interpreting and activating part of the process definition and interacting with the external resources—*invoked web services*—necessary to process the various activities.

A dynamic group of bio-inspired mobile agents *SWEA* (Self-organized Workflow Executor Agent) [13] representing the workflow executors generated from the BPEL workflow specification are initially deployed on the basis of the workflow configuration. A coordination model describes how the generated agents cooperate with each other to reach a choreographic execution. In Sunflower, the coordination model is obtained by the Petri Net (PN) associated with the BPEL program. The PN representation is then structurally decomposed into a set of distributed sub-flow schemas.

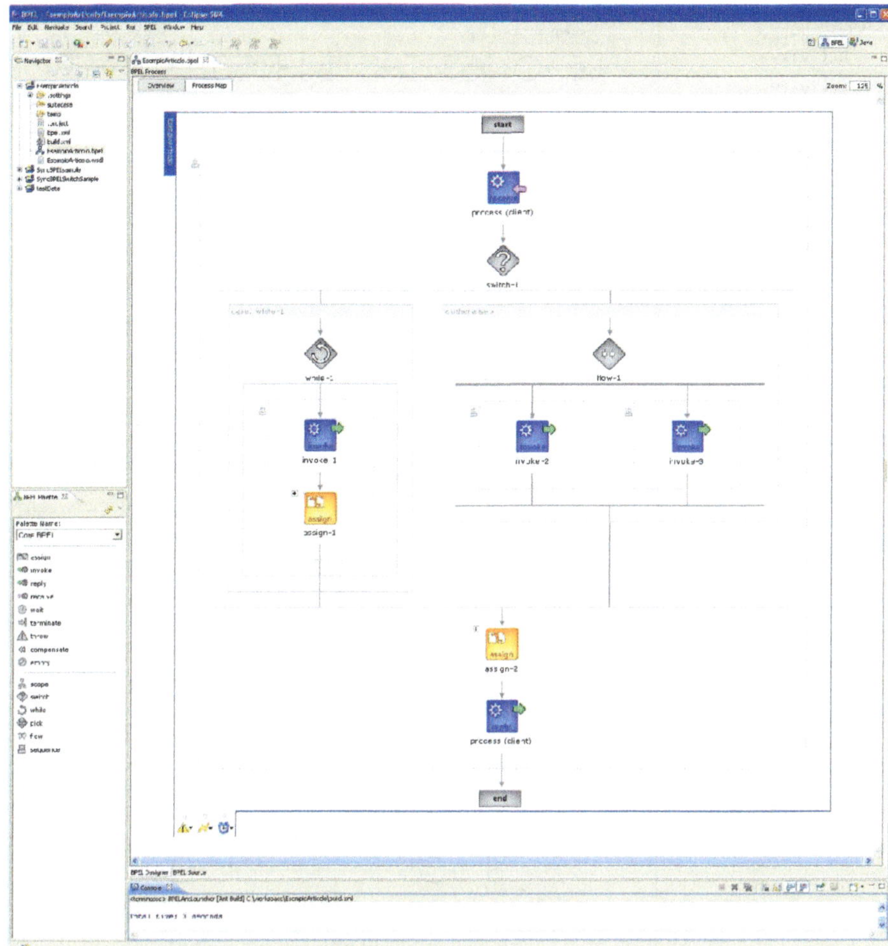

Fig. 3 Example of BPEL workflow (*right*)

3.1 Mapping BPEL Workflows on Petri Nets

A workflow described in BPEL, as shown in Fig. 3, details the flow control and any data dependencies among a collection of Web services being composed. We build every process in a BPEL workflow by plugging language constructs together; we thus can translate each construct of the language into a Petri net (PN). Each primitive or structured activity can be easily modeled as a Petri Net as illustrated in Fig. 4.

BPEL based workflows are converted to a Petri net applying the rules defined by the Van der Aalst methodology [14] that generate a PNs via the repetitive replacement of elemental PNs with other PNs. Figure 5a shows the conversion of the BPEL workflow described in Fig. 3 to a PN form using the replacement property.

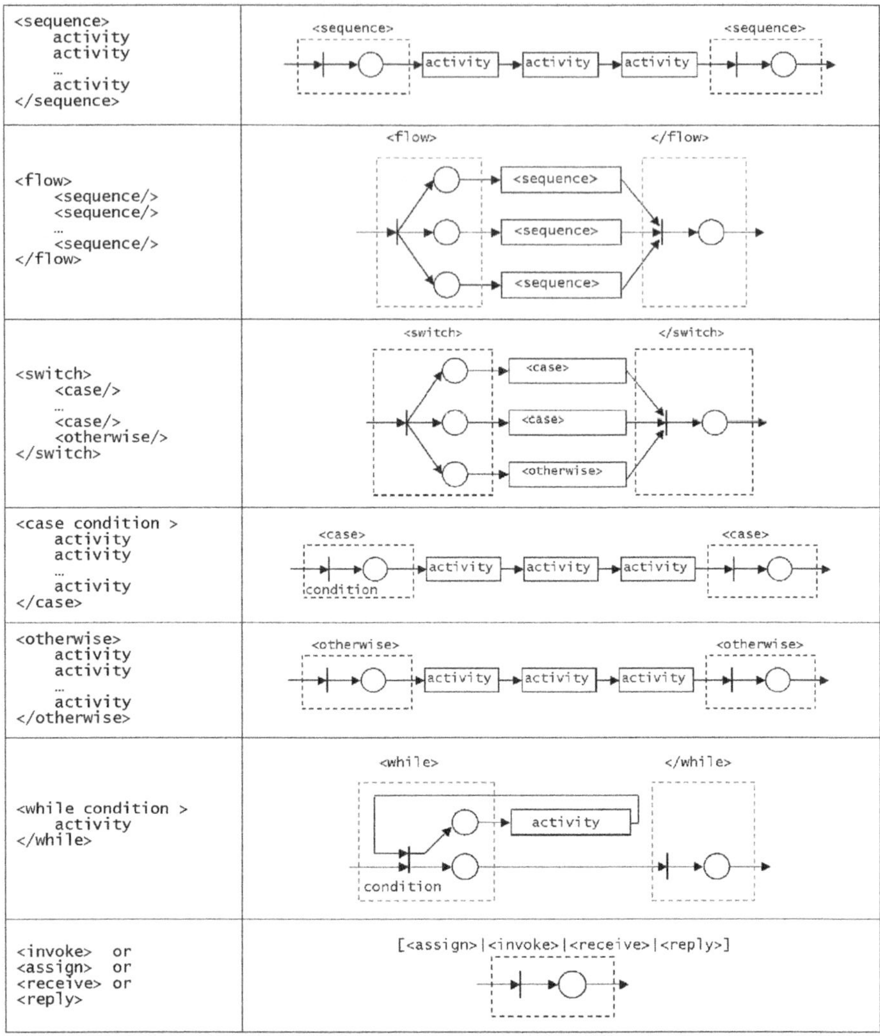

Fig. 4 Example of BPEL constructs (*left*) converted to Petri nets (*right*)

3.2 Petri Nets Partitioning

A workflow written as a single BPEL program must be decomposed in an equivalent set of decentralized processes to set up the choreography model. Our strategy is to construct a PN for the workflow and then apply partitioning rules to operate on such an abstract representation to create the set of cooperating sub-workflows. Our PN partitioning algorithm is based on the idea of merging tasks starting from the *invoke* activities along the control dependence edges.

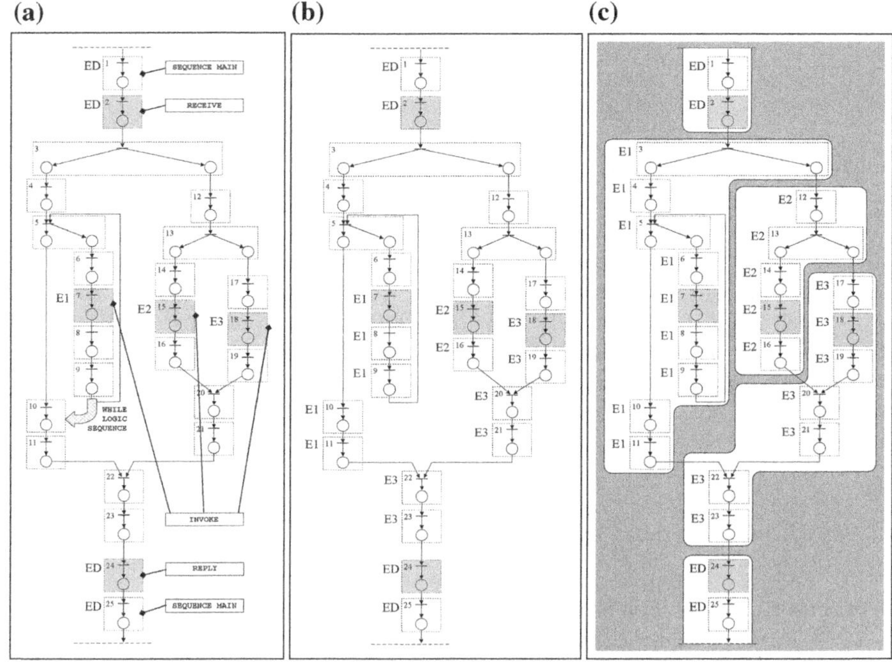

Fig. 5 Petri nets partitioning **a** Petri nets. **b** Top down allocation. **c** Bottom up allocation

An informal description of the partitioning algorithm is as follow:

- The begin and end portions of PN concerning the *main sequence* must be assigned to the same peer, named Peer Collector (PC). *Reply* and *receive* activities must be also executed on the PC.
- The portion of PN concerning the *invoke* is assigned to the peer handling the web service called by the invoke itself.
- All the other constructs are assigned by means of two subsequent visits (*top down* and *bottom up*) of the PN graph. The visits of the PN graph start from the invoke activities. Constructs between two invoke activities can be assigned to one of the two peers where the invokes are executed in order to balance the load.

After the initial allocations, we start with an TopDown visit of the PN graph and then continue with a BottomUp visit. The procedure is as follow:

1. *TopDown initial allocation*: for each PN portion concerning the invoke activity, the label that indicates the peer on which the invoke activity is allocated is propagated to all the *successors*; in the case of controversy (several activities going to the same place), only the *right* label is propagated.

2. *BottomUp final allocation*: for each PN portion concerning the invoke activity, the label of the peer assigned to this activity is propagated to all the *predecessors*; in the case of controversy, only the *left* label is propagated.

In order to better describe the entire process, the PN graph shown in Fig. 5a will be used to illustrate the partitioning procedure. Figure 5b shows how the activities are allocated to the peers through the TopDown procedure. Following the above partitioning algorithm, activities 1, 2 (*main sequence, begin and receive*) and 24, 25 (*reply and main sequence end*) are assigned to the Peer Collector. Then, starting from the invoke activities marked as 7, 15 and 18 activities 7–11, with the E1 label, are assigned to Peer1, activities 15 and 16, with E2 label, are assigned to Peer2 and activities 18–23, with E3 label, are assigned to Peer3. Then, applying the BottomUp procedure, activities 3–6 are assigned to Peer1, 12–14 to Peer2, 17 to Peer3 as shown in Fig. 5c.

3.3 Sunflower Decentralized Execution

On the basis of these schemas, Sunflower enacts the federation of (*SWEA*) agents that has to be executed on the SBE nodes. The decentralized execution of the workflow is coordinate by tokens exchanged among the SBE platforms. Tokens contain the whole execution state, including all data gathered during execution. Each *SWEA* agent performs the portion of workflow assigned and determine which agent should be activated next.

SWEA agents adapt their structure moving over the Internet to position themselves in the nodes with low workload and where the Web services with the best performance are available. The framework provides support for the migration-transparent of the agents and instructs the agents, by a migration policy, to migrate in order to achieve goals like load balancing, performance optimization or guaranteeing QoS.

Sunflower monitors the QoS for Web services by Antares [15] and effectively self-adapts the workflow engines in response to changes in load patterns and server failures. Antares is able to disseminate and reorganize service descriptors by an ant-clustering algorithm and, as a consequence of this, it facilitates and speeds up discovery operations. Based on dynamic service performance evaluation, the services with similar or same metric are gather into clusters by Antares. Scheduling managers make a scheduling decision based on user QoS requirements and information in Antares. All member services in a cluster provide similar or the same QoS after service clustering. Consequently, the task scheduling involves two steps: initial cluster selection from service clusters and further service selection from the selected cluster.

To support workflow adaptation the *SWEA* agents are assisted by routing/scheduling *RA* agents and monitoring/analizing *MA* agents that interact with the Antares information system. The MA agent collects details about the performance metrics and workload of the Web service and when it detects a change, owing to external events, it inserts a new Web service descriptor with the new information in the Antares virtual space and notifies the change to the RA agent. When the RA agent receives a notification about a modification of the class of QoS, it sends a query to Antares to discovery and select a descriptor of an equivalent optimal service. Then, Antares returns a reference to an end point handler for the selected service. Before executing the sub-workflow, the *SWEA* agent contacts the *MA* agent to verify whether the class of QoS of the service to invoke is respected. In the affirmative case, the *SWEA* agent invokes the service and performs the workflow task, otherwise it uses its migration-policy to decide its destination consulting the RA agent. The activities of the *MA* and RA agents are performed continuously.

Sunflower uses the scheduling algorithm executed by RA agents, to make these choices, using information provided by Antares. For each Web service, the RA agent schedules the SWEA agents queued in the local SBE. The scheduled SWEA agent checks whether the QoS of the service relied on the node of the network is less than that required. If affirmative, a request is sent to the Antares registry service to search for an equivalent service to replace. If the service exists and is available, the reference to the service is returned to the RA agent that uses this information to migrate the SWEA agent on the node where the service is localized. Otherwise, if there are no services available an activation request for a new virtual machine is sent to the Cloud [16] provisioner.

Before, a new VM is started the provisioner checks whether there is one VM with the QOS requested already, else a new VM is started. The VM continues the execution of the workflow and before invoking the next Web service on the Cloud the RA agent checks, by querying Antares, whether an equivalent service is available on the Internet. If affirmative, the activation token with the status information is transferred to the SBE node on the network that has the next service to invoke.

4 Integration of RESTful SO in Sunflower

IoT technology emerges from the recent research and technology advancements in the fields of embedded systems and wireless sensor networks [17]. In these contexts, a plethora of electronic objects has been developed that fulfills even more complex requirements. These objects span from simple sensors to more and more flexible and programmable objects. In addition, all the objects should be able to interact with each other and with the *services* on the internet in order to fully accomplish the IoT philosophy. These considerations suggest these objects should be wrapped with a standard well-established interface that also addresses the complex and proactive behavior leading to the concept of *SO*.

REST services could be a way to deal with standardization issues in an easy and lightweight way. Also, REST technology strongly relies on *IP reachability* which is a fundamental concept of IoT technology in the world of IoT researchers.

Our middleware uses a REST interface to wrap SOs and also supplies an *event-driven* engine that properly captures the context modifications in the physical part.

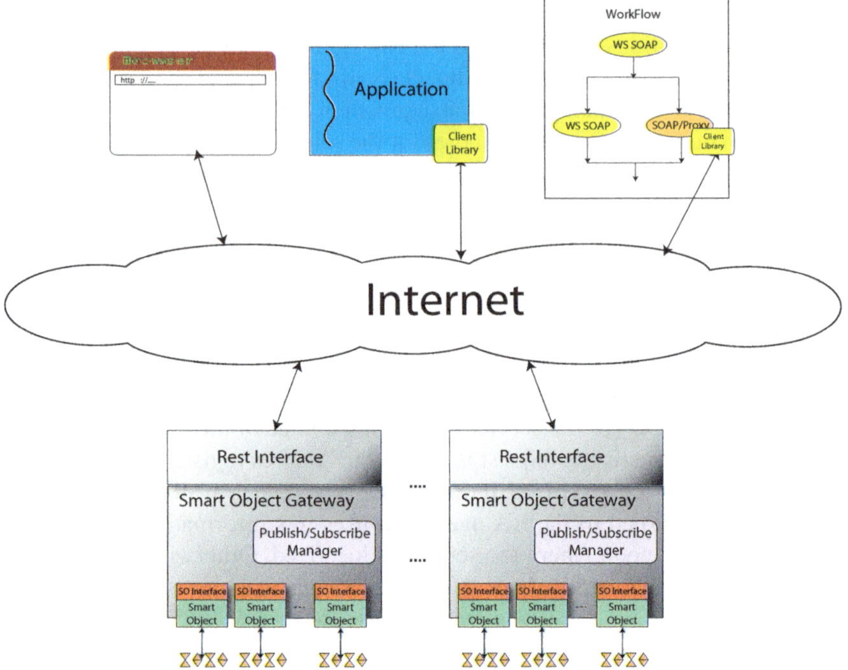

Fig. 6 Middleware architecture for integration of web services and smart objects

Roughly speaking, our middleware permits us to read/write the SOs and define events upon them through a well-established REST interface. Events are defined by *logical rules* submitted to the middleware and exploited in a *publish/subscribe* [18] fashion.

This approach is designed to ensure *ubiquity* and *location transparency* of SOs while fostering a *service-oriented* easy to use development of IoT applications.

As is shown in Fig. 6 our architecture was conceived so that SOs are linked to different computing nodes. Each SO is wrapped in a suitable `Smart Object Interface (SOI)`. All the SOs relative to a computing node, together with their Smart Object Interfaces, are managed by the SOG.

SOG represents the "glue" between the REST part and the SO part of the system. More in particular, SOG is a *singleton* (i.e. one SOG per computing node) and persistent entity of the middleware. Through SOG, REST invocations, which are intrinsically non-persistent and stateless, can access properly to the SOs which are conversely persistent and stateful. In addition, SOG is in charge of managing events defined by rules involving more than one SO; In this case, SOG divides an entire rule into different sub-rules and then assigns each sub-rule to the suitable SO.

As can be seen in Fig. 6 remote accessing of the SOs can be done in different ways: owing to REST protocol, one can access the middleware and relative SOs by using a normal browser, that is, by means of *HTTP protocol*. In a different scenario, one can interact with the middleware using an *ad hoc* client library that hides REST invocation details offering suitable API for developing the application in general purpose programming languages such as: java, c++ and so on.

The third scenario is the one which is more of interest in the context of this work: it foresees full integration between the SO paradigm and web service's orchestration with the addition of an event-driven methodology. Web services orchestration could be realized in different ways, in the context of this work we propose the use of *WS-BPEL/SunFlower* described in the previous sections.

WS-BPEL technology permits web services to be orchestrated through workflow design. Nowadays, many graphical tools exist that allow WS-BPEL workflows to be developed in an easy manner in order to allow people, even with no skills in programming, to create their own applications. WS-BPEL relies on *SOAP* web services rather than on RESTs. Our middleware tackles this issue using *SOAP/REST proxy* services that permit full compatibility and integration.

4.1 SOs Versus Physical Resources

A SO can be composed of just a simple sensor or it can be a more complex object that includes many sensors, many actuators, computational resources like CPU or memory and so on. Examples of complex SOs can be: *smart room*, *smart flat*, *smart building* etc.

In general, SO outputs can be represented by *punctual values* (e.g. the temperature at a given point of a room) or *aggregate values* (e.g. the average of moisture during the last 8 h). Also, the values returned by SOs could be just the measurement of sensors or could be the result of complex computations (e.g. the temperature of a given point of space computed by means of interpolation of the values given by sensors spread across the environment). Furthermore, a SO could supply actuation functionality by changing the environment on the basis of external triggers or internal calculus.

These different kinds of behavior that SO can expose must be reflected in how it is integrated in the middleware. SO is therefore conceived as a complex object that can read and write upon many simple physical resources. More in details, we consider that each SO exposes different *functionalities*. Each functionality can be either sensing or actuating and can be refined by further parameters that dynamically configure it. The previous assumption leads to the definition of *simple physical resource* as the following *triplet*:

$$[SOId, SOFunctionId, Params]$$

where `SOId` uniquely identifies the SO, `SOFunctionId` identifies the specific functionality and `Params` is an ordered set of parameter values that configure the functionality. For example let's consider a *Smart Room* made of sensors for measuring different physical quantities inside a room such as *temperature*, *moisture*, *brightness* and so on. Suppose now you want to read from Smart Room the temperature in a given spatial point of the room. In that case the triplet could be:

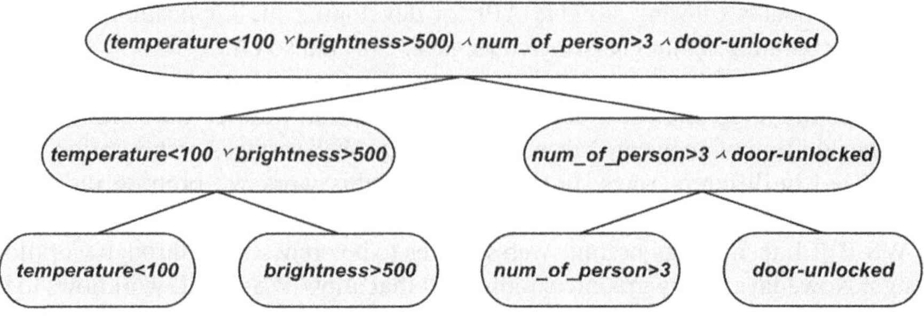

Fig. 7 Example of binary tree of a rule

$$[SmartRoom, temperature, [x, y, z]]$$

where x, y and z are the cartesian coordinate of the point of interest.

4.2 Publish/Subscribe of Events

SOG include a *publish/subscribe* component for managing events in each computing node. Each event is defined by a *logical rule* where one or more SOs could be involved. Each rule is a *logical proposition* in which the *atomic predicates* can be of the following kinds:

simple_physical_resource < threshold (e.g. temperature <300)
simple_physical_resource > threshold
boolean_resource (e.g. the_door_was_unlocked)

Just an example of rule:

(temperature < 100 and brightness > 500) or number_of_ person > 3 *or door_unlocked*

The SOG (specifically its publish/subscribe manager) is in charge to parse the logical rule and generate a *binary tree* made as explained below: each node *N* of the tree corresponds to a logical proposition $N()$. given *L* and *R*, the child nodes of *N*, their associated logical propositions are respectively $L()$ and $R()$ so that it results either *N() = L() or R()* or *N() = L() and R()*. The radix of the tree corresponds to the entire rule while the leaves contain the atomic propositions that SOG considers in order to pass them forward towards the suitable SOs.

An example of a binary tree representation of a composed rule is shown in Fig. 7.

A SO is in charge to establish each time when the assigned atomic propositions are *true* or *false*. The logical proposition of a given node is computed on the basis of the value of its child nodes. The root of the tree is recursively involved by this bottom-up computation. As soon as the value of the root node (i.e. the value of the entire rule) changes SOG notifies all the subscriber.

4.3 Smart Object Interface

The previous sections are focused on supplying a sort of general formalization to SO and their relative functionalities. The effort of formalization is now useful to introduce a well-established interface for SO:

```
interface SmartObjectInterface {
   SOResult checkValue(SOFunctionId functionId, SOFunctionParams
       params);
   SOResult acting(SOFunctionId functionId, SOFunctionParams
       params);
   void setRule(SOFunctionId functionId, SOFunctionParams params,
        Operator operator, SOValue threshold, RuleMatchedListener
        listener);
   void setRule(SOFunctionId functionId, SOFunctionParams params,
        RuleMatchedListener listener);
}
```

Where the `checkValue` is the method for read a particular physical resource, `acting` is the method for performing an operation that produces a change in the smart environment. There are 2 methods `setRule`: the first concerns publishing of threshold based rules while the second is thought for Boolean resources (see Sect. 4.2).

The parameters of the methods follow the previously described logic: `functionId` identifies a functionality that the SO exposes, `params` is an ordered set of parameter values that configure the functionality, `operator` is just the comparative operator to be used for the rule, its value can be either $<$ or $>$, `threshold` is a numeric value intended as the threshold value of the rule. The last parameter of both `setRule` methods is a *listener* object that the SO have to notify when the value of a rule is changed. The involved SO will execute notifications by calling the methods of the `RuleMacthedListener` presented below:

```
interface RuleMatchedListener {
   void ruleMatched();
   void ruleNotMatched();
}
```

For instance, let's consider the previously introduced Smart Room example, if one wants to publish an event that occurs when the *temperature* in the [4,4,5] point of the room's space is less than 27 then the method `setRule` should be called as shown in the following code excerpt:

```
//pseudo code
SmartRoom.setRule(temperature_Id, [4,4,5],Operator.lessThan, 27,
    objectListener);
```

It is worth noting that the `SO Interface` is used only by the SOG while it is completely hidden at application level. SOG is in charge of interacting with the suitable SO on the basis of the application part that, in turn, interacts with SOG through the SOG interface presented in the next section. Finally, all the SOs will have to link itself to the suitable SOG calling the `register` method on the SOG thus supplying its unique *Id*. For example:

```
//pseudo code
SmartObjectGateway.register(SmartObjectId, this);
```

4.4 Gateway Interface

The SOG implements the `GatewayInterface` described below. This interface is exposed outside by means of REST technology. The middleware foresees a suitable *proxy* SOAP web service that executes REST invocations in order to reproduce the `GatewayInterface` in the client side thus permitting fully integration with WS-BPEL/SunFlower workflows.

```
interface GatewayInterface {
   void resourceNaming(String name, SOId soId, SOFunctionId
       functionId, SOFunctionParams params);
   SOResult check(String name);
   SOResult check(String name, SOFunctionParams params);
   SOResult acting(String name);
   SOResult acting(String name, SOFunctionParams params);
   void setRule(Rule rule, String idRule);
   void subscribe(String idRule);
}
```

The method `resourceNaming` assigns an identification `name` to a given resource supplied by a given SO. A resource is a particular instance of a *functionality* of a SO refined by some *parameters*. In other word, a resource is the above-mentioned triplet: [*SOId*, *SOFunctionId*, *Params*]. The `name` assigned to a resource via `resourceNaming` can be used in the other methods in order to simply identify the resource. Furthermore, the identification `name` of a resource is useful to compose the rules in a more human-readable fashion.

The method `check` reads the current value of the resource identified by `name`. `acting` triggers tha actuation operation upon the resource identified by `name`. Both `check` and `acting` methods are of two kinds: the first take only `name` as parameter and refers to the resource as it is previously defined in `resourceNaming`; the second kind, instead, permits to dynamically refine the parameters of the referred resource. The method `setRule` permits a complex rule to be published (e.g. (*tem-*

Fig. 8 Micaz mote and MTS310 sensor board

perature < 100 and brightness >500) or number_of_person > 3 or door_unlocked) and to assigns an id (i.e. `idRule`) useful for subscribing the rule afterwards.

The method `subscribe` permits a previously published rule to be subscribed that is identified by `idRule`. It is worth underlining that the latter method is a blocking method: when it is invoked, the middleware takes care of the event of interest while the execution of `subscribe` is stopped for waiting for the event to occur. This blocking behavior guarantees a correct integration in WS-BPEL/SunFlower workflows. In general, indeed, a workflow has one start point and one end point. Between start and end points there is the entire workflow that can be as complex as required. Nevertheless, it does not have any other entry points of execution except for the start point, so workflow cannot manage asynchronous operations properly because some sort of request/callback mechanism is required to cope with the asynchronous scenario.

5 Example of Usage

In this section we introduce a simple example of usage of the middleware in order to explain in detail how the proposed middleware works. In the example we use *Micaz motes* (Crossbow MPR2400) (see Fig. 8) as reference technology to build SOs.

A Micaz mote is a processor/radio board that run the operating system *TinyOS* [19], which handles power, radio transmission and networking transparent to the user. The Micaz system is widely used in the context of wireless sensors network where multiple motes distributed over a wide area are able to wirelessly transmit their data back to a *base station* attached to a computer.

TinyOS operating system enables Micaz motes to be programmed in *NescC* language supplying the chance to perform even complex elaborations directly inside the mote itself. Each mote can be expanded by attaching it a *sensor board* like MTS310 (see Fig. 8) which includes different kinds of sensing operations concerning physical entities such as *temperature*, *brightness* etc. In addition, each MTS310 sensor board includes 3 *led* that will be used as actuators in the context of our example. Firstly, we need to define the SOs we want, the functionalities they would offer and the meaning of the latter. After, we have to program the Micaz motes properly and develop suitable computer side SOs implementations that interact with motes via base station. In our simple example we define a single SO called `smart micaz` made of 3 Micaz motes. The first 2 motes both have sensing behavior while the third have the role of actuator. More in particular, the foreseen functionality for the first 2 concerns brightness while the third one exposes an actuating functionality that corresponds just to turn on and off a led. Formally, we call `light 1` and `light 2` respectively the sensing functionality of the first 2 motes and we call `actuator` the acting functionality of the third one. On the basis of the just defined functionalities we introduce the *resources* that will be used by the application part. Each resource corresponds to a triplet as explained in the previous sections:

```
light sensor 1 = { smart micaz, light 1, [] }
light sensor 2 = { smart micaz, light 2, [] }
led off = { smart micaz, actuator, [led = 0] }
led on = { smart micaz, actuator, [led = 1] }
```

Now we are free to use the resources as we want, we can read and publish/subscribe events upon `light senor 1` and `light sensor 2` or trigger actuation of `led off` and `led on`. The application we want to develop as example has a straightforward behavior: when the brightness sensed by the first mote (i.e. `light sensor 1`) decreases under a given threshold value, the led upon the third mote shall be turned on (i.e. triggering `led on`) whilst if it is the brightness measured by the second mote (i.e. `light sensor 2`) which decreases under the threshold value, the led upon the third mote shall be turned off (i.e. triggering `led off`). The above-described application is created by setting up a WS-BPEL workflow as shown in Fig. 9.

As can be seen the *flow* construct is used which enables executing commands in parallel. In our example there are two branches (i.e. sub-workflows) executing concurrently, each execution branch is in charge of controlling one of the reading resources (`light sensor 1`, `light sensor 2`) and triggering one of the writing resources (`led on`, `led off`). More in particular, each branch contains a *while* construct that loop infinitely. For each iteration a first proxy web service is called in order to subscribe an event defined by a rule such as `light sensor 1 < 150000` (the `Assign` construct passes the rule as parameter to the `Invoke` construct). The `subscribe` operation waits for the event to occur. When the rule is matched the first proxy web server ends its execution and a second proxy web server is called in order to trigger the actuation part. In that case the `Assign` construct passes `led on` or `led off` as a parameter to the `Invoke` construct.

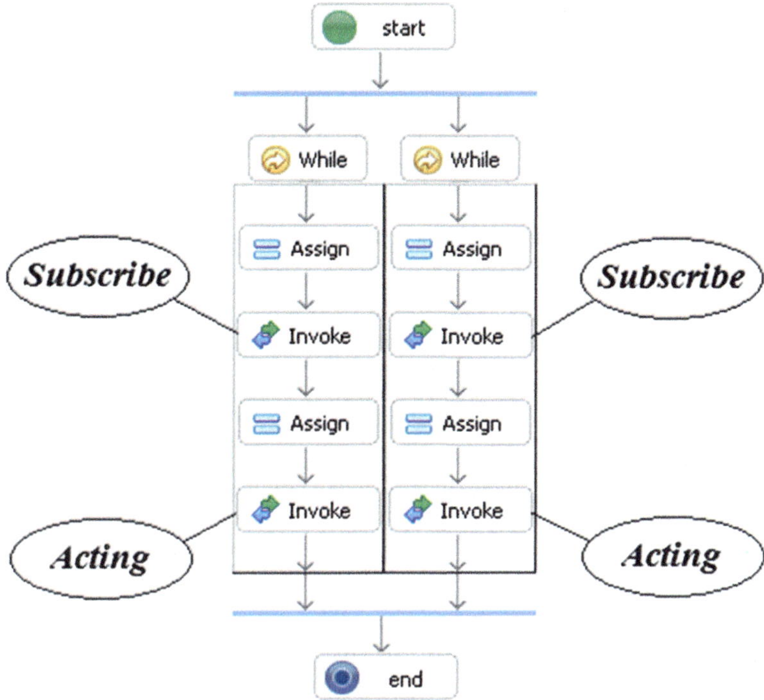

Fig. 9 WS-BPEL workflow

6 Conclusions

This chapter introduces an innovative framework and applications to combine Web service and SOs. The platform designed provides the execution of event-driven choreographic workflows. We present the design requirements and the corresponding architecture with a description of the technologies and platforms we intend to use for the implementation. Practical experience gained with the evaluation and implementation of the architecture demonstrates that it is both feasible and flexible to adapt to a variety of applications and off-the-shelf technologies. Regarding the implementation issues, we have shown that Event-driven Architecture has evolved to Event-driven SOA and this combination may form the foundation of emerging smart systems. Furthermore, we describe an implementation of a simple example of usage of the middleware following the SO's vision. In the near future, we aim to implement and test the proposed solution with the help of a smart room application.

Acknowledgments This work has been partially supported by TETRis—TETRA Innovative Open Source Services, funded by the Italian Government (PON 01-00451).

References

1. ISTAG Working Group Report on We-based Service Industry, February 2008, ftp://ftp.cordis.europa.eu/pub/ist/docs/web-based-service-industry-istag_en.pdf.
2. Alvarez, F., et al. (eds.): The Future Internet—Future Internet Assembly 2012: From Promises to Reality. LNCS, vol. 7281. Springer, Berlin (2012)
3. Atzori, L., Iera, A., Morabito, G.: The internet of things: a survey. Comput. Netw. **54**(15), 2787–2805 (2010). ISSN 1389–1286
4. Kortuem, G., Kawsar, F., Fitton, D., Sundramoorth, V.: Smart objects as building blocks for the Internet of things. IEEE Internet Comput. **14**(1), 44–51 (2010)
5. Fortino, G., Guerrieri, A., Russo, W., Savaglio, C.: Middlewares for Smart Objects and Smart Environments: Overview and Comparison, in Internet of Things based on Smart Objects: technology, middleware and applications. Springer Series on the Internet of Things (2014)
6. Fortino, G., Guerrieri, A., Lacopo, M., Lucia, M., Russo W.: An agent-based middleware for cooperating smart objects. In: Highlights on Practical Applications of Agents and Multi-Agent Systems, Communications in Comp. and Inform. Science (CCIS), vol. 365, pp. 387–398. Springer, Berlin (2013)
7. Fortino, G., Guerrieri, A., Russo, W.: Agent-oriented smart objects development. In: Proceedings of IEEE 16th International Conference on Computer Supported Cooperative Work in Design (CSCWD), pp. 907–912 (2012)
8. Teixeira, T., Hachem, S., Issarny, V., Georgantas, N.: Service oriented middleware for the internet of things: a perspective. In: Proceeding ServiceWave'11 Proceedings of the 4th European Conference on Towards a Service-Based Internet, pp. 220–229. Springer, Berlin (2011)
9. Barker, A., Besana, P., Robertson, D., Weissman, J.B. : The benefits of service choreography for data-intensive computing. In: CLADE '09 Proceedings of the 7th international workshop on Challenges of Large Applications in Distributed Environments. ACM, New York, pp. 1–10 (2009)
10. Papuzzo, G., Spezzano, G.: Processing applications composed of web/grid services by distributed autonomic and self-organizing workflow engines. In: Chapman, B., Desprez, F., Joubert, G.R., Lichnewsky, A., Peters, F., Priol, T. (eds.) Parallel Computing: From Multicores and GPU's to Petascale. Advances in Parallel Computing. IOS Press, vol. 19, pp. 195–204 (2010)
11. Rahman, M., Buyya, R.: An autonomic workflow management system for global grids. In: Proceedings of the 8th IEEE International Symposium on Cluster Computing and the Grid (CCGrid 2008), IEEE CS Press, pp. 578–583 (2008)
12. Alves A., et al.: Web Services Business Process Execution Language (WS-BPEL) 2.0, OASIS, August 2006, http://www.oasis-open.org/committees/wsbpel, accessed 12 Feb 2010
13. Brazier, F.M.T., Kephart, J.O., Van Dyke Parunak, H., Huhns, M.N.: Agents and service-oriented computing for autonomic computing: a research agenda. IEEE Internet Comput. **13**(3), 82–87 (2009)
14. van der Aalst, W., van Hee, K.M.: Workflow Management: Models, Methods, and Systems. MIT Press, Cambridge (2002)
15. Forestiero, A., Mastroianni, C., Spezzano, G.: Antares: an ant-inspired P2P information system for a self-structured grid. In: BIONETICS 2007–2nd International Conference on Bio-Inspired Models of Network, Information, and Computing Systems, Budapest, Hungary, 2007
16. Buyya R., Yeo C.S., Venugopal S.: Market-oriented cloud computing: vision, hype, and reality for delivering IT services as computing utilities. In: Proceedings of the 10th IEEE International Conference on High Performance Computing and Communications (HPCC 2008), IEEE CS Press, Los Alamitos, CA, USA, Sept. 25–27, 2008, Dalian, China. - Keynote Paper, 2008
17. Raghavendra, C.S., Sivalingam, K.M., Znati, T. (eds.): Wireless Sensor Networks. Springer, Berlin (2004)
18. Eugster, P., Felber, P.A., Guerraoui, R., Kermarrec, A.: The many faces of publish/subscribe. J. ACM Comput. Surv. (CSUR) **35**(2), 114–131 (2003)
19. Levis, P., Madden, S., Polastre, J., Szewczyk, R., Whitehouse, K.: TinyOS: An Operating System for Sensor Networks. Springer, Berlin (2005)

CO-Based Outdoor Smart Lighting for Energy Aware Factory

Anna Florea, Ahmed Farahat, Corina Postelnicu, Jose L. Martinez Lastra and Francisco J. Azcondo Sánchez

Abstract Energy awareness together with holistic perception of consumption processes are one of the main factors contributing to efficient and sustainable performance of such complex systems as buildings, cities, and factories. Availability of relevant data and possibility for cross-domain integration become minimum requirements defining the success of the implementation. Emergence of cooperating smart objects, resulting from evolution in IoT and embedded devices, helps achieving both energy awareness and efficiency by offering possibility of sensing and acting over complex environments and overcome challenges associated with cross-domain integration. This chapter describes smart lighting application for the industrial outdoor environment implemented using cooperating objects featuring Semantic Web Service middleware. Presented use-case considers the university campus area comprising multipurpose outdoor area and neighbouring industrial laboratory facilities. The application is aiming efficient use of energy and possibility for integration with relevant industrial systems.

Keywords Smart cooperating objects · Embedded devices · Web services · Energy efficiency · Smart lighting

A. Florea · A. Farahat · C. Postelnicu · J. L. Martinez Lastra (✉)
Tampere University of Technology, Tampere, Finland
e-mail: anna.florea@tut.fi

C. Postelnicu
e-mail: corina.postelnicu@tut.fi

J. L. Martinez Lastra
e-mail: jose.lastra@tut.fi

F. J. Azcondo Sánchez
University of Cantabria, Santander, Spain
e-mail: javier.azcondo@unican.es

G. Fortino and P. Trunfio (eds.), *Internet of Things Based on Smart Objects*, Internet of Things, DOI: 10.1007/978-3-319-00491-4_4,

1 Introduction

Energy awareness and efficiency is an important application domain for modern Information and Communication Technologies (ICT) ranging from enterprise integration to IoT and smart cooperating objects. Together with holistic perception of consumption processes, it supports efficient and sustainable performance of such complex systems as buildings, factories and cities. The design success is heavily dependent on the availability of relevant data and possibility for cross-domain integration, which can be realised nowadays with help of the recent advancements in IoT technologies.

Early smart applications targeted stand-alone appliances within small living and/or work spaces. With the adoption of more mature technology, it became possible to embed smart applications in more complex environments, exhibiting different levels of demands and requirements depending on the target domain (e.g. cities [8], factories [2], and lower level industrial environments [26]).

Smart applications target enhanced user experience while facilitating efficient use of resources. There are common challenges across all the application domains that complicate the achievement of these objectives. The challenges include, but are not limited to: multitude of purposes for which the same environment may be used, big amounts of users with different profiles, and dynamics of ambience.

WSN and IoT were successfully applied to implement the early applications for smart environments [23, 24], and continued to evolve driven by newly appeared challenges. The concept of cooperating objects (CO) [14] and smart cooperating objets(SO) [6] bear on the same technological base, as IoT and WSN and is perceived as foundation for the future IoT [10]. These approaches enable creation of sustainable smart solutions for complex applications.

This chapter describes an approach to implementation of smart applications in a multipurpose environment following the cooperating objects paradigm aiming at increased energy awareness, reduced power consumption and enhanced user experience. The use case presented is a smart lighting application for outdoor docking environment at a university campus next to the industrial laboratory. In addition to primary objectives, the solution is intended for the integration with industrial systems located inside the building. The paper is structured as follows: Sect. 2 provides the research background discussing the technological considerations for smart lighting applications; Sect. 3 describes the implemented smart lighting solution; Sect. 4 draws the conclusions and outlines the future work.

2 Technological Considerations for Smart Lighting Applications

2.1 Illumination

Lighting conditions have strong impact on everyday life and individual work performance [7, 15]. Illumination accounts for 5 to 10 % of total energy consumption on the planet [17], with lighting systems presenting huge potential for energy savings

[25]. It is therefore of crucial importance that smart lighting applications should aim efficient resource usage. Most of the savings can be achieved via suitable (multiple type) control strategies [22], that have proven so far more effective than simple personal, institution, occupancy and day-lighting driven control [25].

Recommended illumination levels (as produced by the Illumination Engineering Society) vary from 100 lux in the warehouse areas to 5000 lux for fine inspection operations [13]. In multipurpose environments, compliance with levels tailored for the specific needs of one working environment at hand is achievable via a control strategy allowing to switch between pre-set lighting modes correlated to specific user needs.

As far as energy consumption of lighting solution is concerned there are a number of aspects to be taken into account. The luminous intensity drops rapidly as distance from the light source to the observer increases. Because of the non-linear nature of this dependency, implications of different lighting modes on energy consumption are not as straight forward as it may be initially expected [13]. Although there are simulation tools available allowing to estimate the energy consumption of lighting applications, it has been found that simulations tend to overestimate savings [25]. Therefore, real measurements are needed in order to evaluate the energy efficiency of a lighting solution in place.

The identified challenges may be addressed by considering LEDs over conventional light sources and by implementing smart lighting control customized for the specifics of the environment. Numerous lighting solutions targeting energy efficient performance were developed in previous decade, actively exploiting low consuming light sources, control techniques targeting low energy consumption [1] and the combination of both [19]. The challenges of adopting the best practices of smart lighting solutions are related to the fact, that each application of this kind must be tailored to the needs of the dedicated users and consider peculiarities of the specific environment. On the other hand most of the recent solutions rest on same core architectural paradigms, discussed in the following section.

2.2 Architectural Paradigms

WSN technology is used for many different applications, including structural health care monitoring, habitat monitoring, fire detection or ambient intelligence [2]. A WSN or WASN (wireless sensor and actuator network) is composed of a set of nodes distributed over an area of interest. The nodes are able to sense, process, drive, store and communicate. The network produces large amounts of raw data which then sent to the central server via sink nodes. Some variations of the concept were proposed by the research community looking to enhance either the autonomy of the network (Autonomic Sensor Networks [12]) or data processing and reuse through dynamic tagging of semantic information (Semantic Sensor Networks [16]).

Leveraging RFID and WSN, the IoT aims to break the border between physical and virtual reality through the creation of objects with a virtual representation, which can be integrated into a network of a global scale to interact with each other.

A generic definition is formulated in [3] as follows:

> The main tenet of the IoT is extension of Internet into physical world, to involve interaction with a physical entity in the ambient environment.

The entity may be an entity, a device (the means of integration of the entity with the virtual world), a resource (the software component), or service (defines standardized interfaces and processes for interaction with entities).

There are many definitions of the IoT proposed [20], with a definition focus shifting in time from the objects themselves to their communication capabilities. The notion of *cooperative IoT* can also be found in the literature [9]. Despite the focus shift, there are three core features mentioned across all definitions: (1) global scale of the application, (2) big amounts of devices, (3) heterogeneity of the devices.

Succeeding the IoT, the notion of cooperating objects emerged initially defined at the abstract level in [4] in the following way:

> ... a Cooperating Object is a single entity or a collection of entities consisting of: Sensors, controllers (information processors), actuators or cooperating objects that communicate with each other and are able to achieve, more or less autonomously, a common goal.

While the components of an object are provided in the definition above, the term cooperation does require further clarification.

In [14] cooperation is defined as

> the ability of individual entities or objects to use communication as well as dynamic and loose federation to jointly strive to reach a common goal, which will typically be a goal in sensing or control.

A similar explanation of cooperation is given in [9].

Dynamic cooperation relying on complex messaging patterns with nested messaging threads is highlighted as the minimum technology needs to make object integration combining both visions a reality [11, 21].

The heterogeneity of devices is resolved by using semantic web service (SWS) middleware for in embedded devices [18]. This enables CO to be used for complex cross-domain applications, e.g. smart grid enabling smart houses to communicate with energy providers, marketplaces, alternative energy sources, etc. [9].

Further enhanced connected objects, characterised by ability to intelligently interpret the information are referred as smart cooperating objects (SO) [5]. The key enabler of the paradigm is seen to be the agent technologies due to the matching characteristics of intelligent agents and smart objects (i.e. autonomy, scalability, responsiveness and pro-activeness) [6].

WSN and the IoT are paradigms that provide tools and methods for implementations of the solutions for complex smart environments. The approaches are often used side by side complementing each other in order to fulfil all the needs of the unconventional use-cases. This becomes possible due to the similarity of the technological base, which converges into the notion of CO.

3 Case Study: Smart Lighting Application

The solution described in this chapter was designed to provide appropriate illumination for the multipurpose outdoor environment in a specific utilisation mode using low amounts of electrical power.

The section is split in five parts, dedicated to the description of the testbed (Sect. 3.1), analysis of the utilisation modes of the area (Sect. 3.2), description of the designed architecture (Sect. 3.3), implementation and testing (Sect. 3.4), and the opportunities for integration with other industrial applications (Sect. 3.5).

3.1 Target Environment

The proposed solution is intended for the backyard area auxiliary to one of the buildings of Tampere University of Technology (Tampere, Finland) showed in Fig. 1. The area is used for a variety of purposes, including:

- Students and personnel everyday access to the building via two entrance doors.
- Load/unload of material/equipment to/from trucks via two additional dedicated doors.
- Parking purposes (there are several parking spaces in the area).

The zone is illuminated with four lamps, which are turned on and off following the work time schedule and security guidelines (i.e. some of the lamps are on during the night time to provide minimum illumination to the area); furthermore, the lamps are always on during the darkest period of winter. The existing operational pattern fulfils the basic need for lighting, but does not consider such important aspects as current utilisation mode of the area and nature of the environment hosted by the building.

As it was previously mentioned, there are different types of actors attending the area: students, research and support personnel, and vehicles of various scales. Each of the actors has own purpose when visiting the area, thus lighting conditions tailored for particular utilisation scenario could facilitate the goal achievement and offer better user experience to the users of the area. The part of the building facing the area considerably differs from average study blocks, being more similar to industrial environment, rather than administrative building. The area is actively used as a docking station, and preparations for load and unload operations could become easier if the lighting was automatically adjusted to the activity (i.e. proper lamps were turned to the need intensity to illuminate the working area).

In addition to the above mentioned problems, the existing lighting installation lacks energy efficiency due to the type of lamps used, applied control strategy and lack of dimming capabilities. These obstacles are easily overcome by migration to LED lamps with drivers offering dimming functionality, which is expected to turn into even bigger savings as cold climate prevents overheating of the diodes. Furthermore availability of information on energy consumption patterns, achieved through the site monitoring, can support elaboration of campus-scale energy efficient solution.

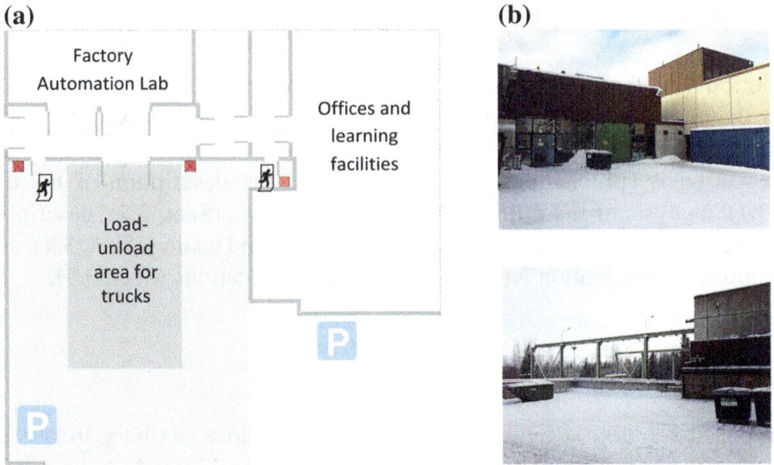

Fig. 1 Testbed area: **a** building and yard plan; **b** view to the front wall with access doors and legacy lamps mounted (*top*), view to the parking area (*bottom*)

3.2 Defining the Representation of the Area State

The key to the improved user experience lays in the knowledge on the current status of the area. Several criteria were considered during the study of the presented multipurpose environment. The most descriptive parameters, selected for the implementation are:

Users present indicate both the fact of presence and the category of users.
Weather conditions focusing on the climate dimensions influencing the visibility.
Illumination level Illumination level provided by the natural conditions.

The notion of profile (P) was introduced in order to combine multiple criteria in one parameter. Profile allows to uniquely identify the superposition of the dimensions as shown in Table 1. Each profile is marked with unique identifier P_{ijk} where indexes stand for one of the alternative values of the profile dimensions (e.g. P_{111} corresponds to a situation when there are people in the area, the sky is clear and it is bright outside).

Each profile is mapped to a specific *lighting scene* (S). It is a collection of operating modes to be assigned to each lamp on-site in order to provide the desired lighting conditions. The range of operating modes varies depending on the lamp and may consist of either *on* and *off* modes, or include a set of intermediate stages if dimming features are available.

Table 2 illustrates mapping between profiles and lighting scenes. It is important to realize, that total amount of lighting scenes is smaller than the amount of profiles, because same combination of lamp operating modes may apply to more than one profile. Therefore the approach results in a reasonable number of lighting scenes to be set up (in the first round of scene design 20 lighting scenes were identified for 32 profiles). The provided tabular representations were used as input for the algorithm

Table 1 Profile dimensions

Illumination level	Users present	Weather conditions			
		Clear	Mist	Rain	Snow
Above threshold	Personnel	P_{111}	P_{112}	P_{113}	P_{114}
	Truck	P_{121}	P_{122}	P_{123}	P_{124}
	Both	P_{131}	P_{132}	P_{133}	P_{134}
	None	P_{141}	P_{142}	P_{143}	P_{144}
Below threshold	...				

Table 2 Profile to lighting scene mapping

P	Lamp operating modes (% of total power)				S
	L_1	L_2	L_3	L_4	
P_{111}	0	0	0	0	S_1
...			...		...
P_{244}	30	50	50	0	S_m

design, helping to identify possible future variations related to change in amounts of lamps and profile dimensions, as hard-coded implementation prevents the scalability of the solution.

3.3 Architecture

The application was designed to serve the two purposes: provide users of the area with lighting conditions adjusted to their needs, provide detailed information about the energy consumed by the installation. The first objective can be easily achieved through sensing of the environmental conditions and user detection and consequent mapping of the detected profile to the required lighting scene. The second objective puts requirement for synchronization of the measurements recordings with the profile changes and raises the question about the degree of granularity of energy measurements. Considering the need to investigate the energy consumption patterns and obtain detailed information about the performance of the updated lighting system, it has been decided to measure consumption of each individual lamp block installed.

The designed architecture is shown in Fig. 2. Due to the small scale of the target area, only four proximity sensors (denoted as PS1–PS4 in the figure) are needed for user detection. Sensors allow detecting the direction from which a user is approaching the area as well as distinguishing between trucks and people. Additionally a Temperature-Humidity-Light (THL) wireless sensor nodes are needed to sense illumination level and weather conditions. The complete information about weather conditions is formed by data from THL sensor and weather web service, which receives full weather profile of the location from a third party weather service (Weather-Yahoo![1]).

[1] http://developer.yahoo.com/weather/.

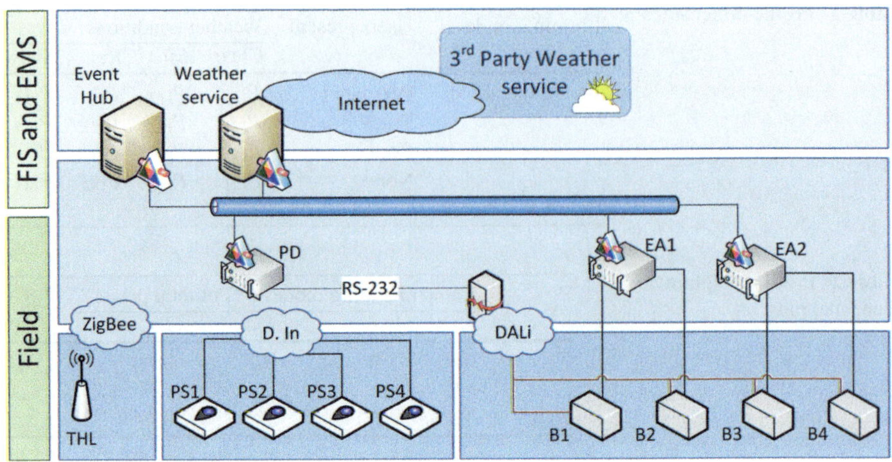

Fig. 2 Smart lighting application architecture

The command application is distributed over three embedded devices, supporting SWS. The first node (denoted as PD in the figure) hosts the main application, receives input data and communicates commands to the lamps. The other two devices (EA1 and EA2) are energy analysers; they are intended to measure the individual energy consumption of the lamps installed. The required amount of energy analysers depends on the amount of actuators (i.e. lamp drivers) and the required granularity of measurements. Each actuator in turn may serve several lamps. Its capacity is limited by the total power of the load attached.

Targeting detailed measurements of energy consumption, each of the lamps is provided with a dedicated actuator. The main controller communicates with individual luminaires via a gateway, which transforms the messages received via serial port into native Digital Lighting Addressable Interface (DALI) messages, understood by the lamps' drivers.

Finally, all three command devices feed the events reporting measurements and status change for further archiving or use in adjacent systems.

From the perspective of the visibility of the components, the overall solution is composed of three layers: distributed sensing and actuating devices in the field, networked cooperating objects hosting the distributed control solution, and external components. The field deployment is aware only of immediately connected control devices; the external components have access to site-scale network (Event Hub) and Internet; the control devices bring together information from the field and the external sources to execute the control and expose field information to the outside applications via web-service interface.

Fig. 3 Hardware components of the implementation: LED lamp with integrated driver, proximity sensor for outdoor use, wireless THL sensor node, two devices S1000 with wireless communication and energy analyser expansion modules

Table 3 Atomic program categories

Service	Monitoring	Control
Variables value updates from device I/Os	Regular energy measurements	Profile change notifications
Variables value updates from external service data	Over-threshold energy measurements	Consumption status acquisition
Scale transformation	On-demand energy measurements	Lighting scene set-up command
Local KPI computation		

3.4 Implementation and Testing

Devices used for the pilot implementation, except the RS-232/DALI gateway, are shown in Fig. 3. Devices hosting the command logic are three S1000 RTU modules: one with extension for wireless communication (PD) and two with E10 expansion modules for monitoring of energy consumption (EA1 and EA2). The outputs of the proximity sensors are wired to the digital inputs of the PD, and W-Z-THL sensors are communicating the measurements via ZigBee PRO protocol. Each of the energy analyser allows to measure energy consumption and related parameters for three phases. In presented scenario, every phase is assigned to particular driver and each analyser is in charge of two drivers, helping to distribute evenly the processing load. The drivers are integrated in the luminaires and are located behind the light sources.

The command functionality is realized through a set of distributed control and monitoring applications. Programs run in S1000 nodes are implemented in Structured Text (ST) language of IEC 61131-3 standard. The weather service is implemented in Java programming language using the Spring framework. The application uses the Weather-Yahoo! API to obtain weather information and interpret it in terms of visibility characteristics defined in Table 1. This information complements the values obtained from THL sensors and helps their adequate interpretation.

The application is implemented in set of atomic programs serving different aspects of the overall solution (Table 3). Service applications are in charge of timely update of variables with data coming from various I/Os and do not communicate with other control devices. Programs contributing to the monitoring acquire information following different time patterns and communicate with other programs triggering measurements via WS interface. Control program set uses results of service applications to manage monitoring and derive control commands for setting up the required lighting scene.

Control and monitoring functionality is implemented in parallel and there are two processes executed in parallel in the devices. Sequence diagram in Fig. 4 illustrates messaging patterns of two possible scenarios: profile change and energy consumption measurements.

When main application receives sensor data it identifies the corresponding profile. In order to avoid big amounts of nested `IF` statements, the profile ID is computed as function of tree profile variables. Then the associated lighting scene is identified. If the computed scene is different from the current one, main application sends a series of messages to the drivers via gateway in order to set up the new scene. Then a notification sent to the energy analysers about the profile change. This message triggers response messages from analysers, containing data on energy consumption. Main application receives data from the analysers and composes a message to be sent to the data acquisition application. It is important to obtain the energy performance information from all the lamps when the profile changes. Therefore, when the first analyser receives the request from the main application, it updates own knowledge about the profile and composes a message containing requested energy data. But, instead of sending the data to the requesting device, it passes the request together with own reply to the second analyser. The second device also updates its profile data and ads requested energy information to the message received from the first device. Finally the information is passed to the main application, where it is used to compose the message to be sent to the adjacent systems via the Event Hub.

Besides the scenario described above, energy analysers perform regular measurements of energy consumption and related parameters. The frequency of measurements is dependent on the current profile. In order to reduce amount of traffic and detect abnormal consumption patterns, measured data are sent to the data acquisition application in the two following cases:

- The nominal time interval defined for the given profile has elapsed;
- The amount of total energy consumed has increased for a value bigger than the threshold defined.

The above mentioned criteria are applied to each phase separately, as different lamps connected to same analyser may be set to different operating modes in certain lighting scenes.

The control box, containing the S1000 modules, is located inside the building, while the lamps and sensors are located outside, which complicates the balancing and commissioning. The correctness of the logic and message flows has been verified in the testbed as depicted in Fig. 5. User presence was simulated by changing the status of digital inputs of the PH device and the light intensity was illustrated through the

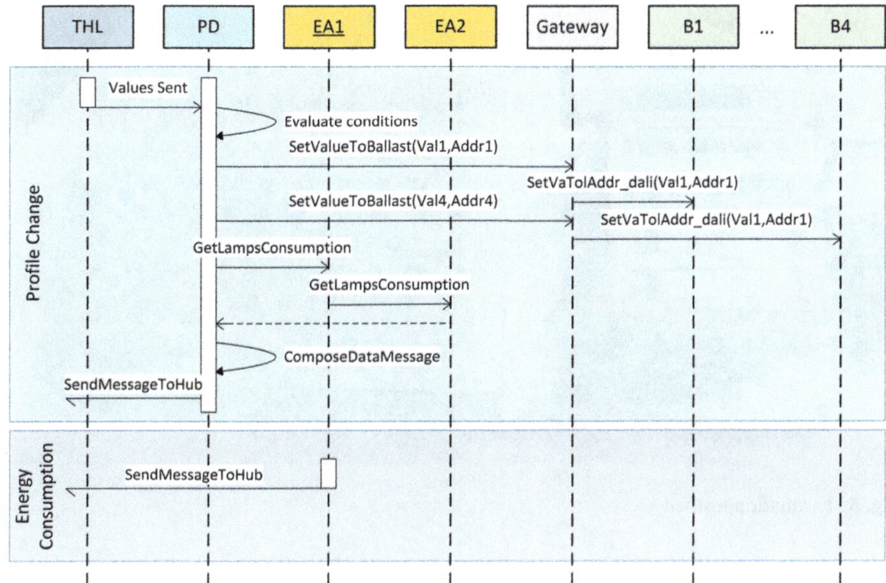

Fig. 4 Sequence diagrams of possible operation scenarios

amount of digital outputs turning on in the same device. Dedicated digital outputs of energy analysers were used to track the message flows, both between the devices and to the external applications.

3.5 Integration with Industrial Applications

Support of SWS at the device level enables direct integration of the solution with the other applications requiring the information produced by the smart lighting application. Possible integration scenarios are considered in this section. During its execution, the designed smart lighting application produces data sent to the data acquisition application for storing. However some of this information may be used in other real-time monitoring and control applications.

As most of the data generated by the application relates to energy consumption of the drivers and their operating modes, it can be included in energy monitoring applications as a separate set of parameters as well as a component of a composite key performance indicator (KPI) e.g. total energy consumption of the site. For the discussed case-study the site consists of the testbed depicted in Fig. 1 and the neighbouring facilities of the Factory Automation Systems and Technology laboratory hosting the production line (Fig. 6).

Fig. 5 Evaluation testbed

The line consists of 10 manufacturing cells, each containing at minimum one robot and a conveyor system. The line is capable of drawing 729 different layouts of mobile phones, using different combinations of frame, keyboard, and screen types. Cell 1 is in charge with determining whether incoming pallets are occupied with finished products and they need unload to be performed on them, or they need further circulation in the line. Quality inspection takes place also here via a machine. The buffer is implemented at Cell 7.

The integration becomes possible due to the availability of the Event Hub (see Fig. 2), receiving the WS messages from the command devices and directing them to the subscribed applications. A client application was developed to receive the messages from the smart lighting application and store it in the MySQL database (DB). It subscribes to for the required messages from the hub, parses them following the information on the system configuration contained in the dedicated XML file and stores information in the database using Hibernate library to interface the DB.

From the perspective of the aims of the lighting application, integration with shop-floor systems is required for truly holistic control strategy both in the manufacturing site and related outdoor area, as well as improved user experience. Extending the described setup to a bigger scale, data received from the proximity sensors may be used to create notifications for personnel and machines about readiness of the docking area for load and unload operations, avoiding centralised control and allowing emergent behaviour of the system. Smart lighting application, in turn, could benefit from receiving of information from the above mentioned applications or the line controllers via the event hub. This opportunity enables implementation of light control scenarios driven by the status of the production environment, e.g. setting up lighting scene required for loading and unloading operation as soon as both truck and line are ready for the process to be started, avoiding influence of human factor in the

Fig. 6 Production line

environment adjustment process which can be regulated by safety and security policies. It could also save a lot of time for the personnel, especially when needed lighting conditions are provided by a big amount of lamps with individual manual switches.

4 Conclusions and Future Work

Notions of CO and SO rely on the technological base similar to one of IoT and WSN, and comprises features allowing taking applications for smart environments to a new level. It enables creation of sustainable smart solutions for such complex environments as smart grid, urban transportation systems, etc.

The paper presents an implementation of smart lighting application in a multipurpose environment following the CO's vision to overall energy awareness of the site. The use case presented is a smart lighting application for outdoor docking environment at a university campus. The information about the status of the test-bed is acquired via a set of wired and wireless sensors and the core functionality is implemented in three networked embedded devices featuring SWS middleware. The application evaluates status of the environment, and manipulates the lamps' drivers in order to set up proper illumination. It also measures the energy consumption of individual drivers allowing evaluation of control strategy from energy efficiency perspective.

Future work will concentrate on such incremental improvements of the solution as fine-tuning of the lighting scenes and optimisation of control application, optimized device and application configuration, as well as its further integration with tools for holistic energy management.

References

1. Bhardwaj, S., Syed, A.A., Ozcelebi, T., Lukkien, J.: Power-managed smart lighting using a semantic interoperability architecture. IEEE Trans. Consumer Electron. **57**(2), 420–427 (2011)
2. da Rocha, A.R., Delicato, F. C., de Souza, J.N., Gomes, D. G., Pirmez, L.: A semantic middleware for autonomic wireless sensor networks. WMUPS'09, June 16, 2009, Dublin, Ireland
3. De Barnaghi, S.P., Bauer, M., Meissner, S.: Service modelling for the internet of things. In: Federated Conference on Computer Science and Information Systems (FedCSIS), 2011, pp. 949–955, 18–21 September 2011
4. Embedded WiSeNts Roadmap, available online: http://www.embeddedwisents.org/dissemination/roadmap.html
5. Fortino, G., Guerrieri, A., Russo, W.: Agent-oriented smart objects development. In: Proceedings of IEEE 16th International Conference on Computer Supported Cooperative Work in Design (CSCWD), pp. 907–912 (2012)
6. Fortino, G., Guerrieri, A., Lacopo, M., Lucia, M., Russo, W.: An agent-based middleware for cooperating smart objects. In: Highlights on Practical Applications of Agents and Multi-Agent Systems, Communications in Comp. and Inform. Science (CCIS), vol. 365, pp. 387–398. Springer, Berlin (2013)
7. Hedge, A., Sims, W.R., Becker, F.D.: Lighting the computerized office. Presentation at the Human Factors Society, October 1989
8. Hernandez-Munoz, J.M., Vercher, J.B., Muñoz, L., Galache, J.A., Presser, M., Hernández Gómez, L.A., Pettersson, J.: Smart cities at the forefront of the future internet. In: The future Internet, pp. 447–462. Springer, Berlin (2011)
9. Karnouskos, S.: The cooperative Internet of things enabled smart grid. In: Proceedings of the 14th IEEE International Symposium on Consumer Electronics (ISCE2010), June 7–10. Braunschweig, Germany (2010)
10. Karnouskos, S., Villaseñor, V., Handte, M., Marron, P.J.: Ubiquitous Integration of Cooperating Objects. Available on-line at: https://e67d01f2-a-62cb3a1a-s-sites.googlegroups.com/site/handteorg/about/papers/2011-IJNGC-Ubiquitous-Integration-of-Cooperating-Objects.pdf. Cited 20 Mar 2013
11. Karnouskos, S., Villaseñor-Herrera, V., Haroon, M., Handte, M., Marrn, P.J.: Requirement considerations for ubiquitous integration of cooperating objects. In: 4th IFIP International Conference on New Technologies, Mobility and Security (NTMS), pp. 1–5, 7–10 February 2011
12. Kephart, J.O., Chess, D.: The vision of autonomic computing. In: IEEE Computer Magazine (2003)
13. Light up the visual Factory. Lighting Solutions for Lean Manufacturing. White paper. Available on-line at: http://www.controleng.com/single-article/sponsored-white-paper-light-up-the-visual-factory/0b99905f978febb7bd0fe374f03be263.html. Cited 16 January 2013
14. Marron, P.J., Minder, D.: European research on cooperating objects. In: 6th Annual IEEE Communications Society Conference on Sensor, Mesh and Ad Hoc Communications and Networks Workshops, 2009. SECON Workshops '09, pp. 1–3, 22–26 June 2009
15. Melchore, J.A.: Sound practices for consistent human visual inspection. AAPS PharmSciTech **12**(1), 215–221 (2011)
16. Ni, L.M., Zhu, Y., Ma, J., Li, M., Luo, Q., Liu, Y., Cheung, S.C., Yang, Q.: Semantic sensor net: an extensible framework. In: Proceedings of ICCNMC. ICCNMC (2005)

17. Nisson N., Wilson, A.: Virginia Energy Savers Handbook, 3rd ed, chap. 8. Virginia Department of Mines, Minerals and Energy, Richlands (2008)
18. Ramos, A.V., Delamer, I.M., Lastra, J.L.M.: Embedded service oriented monitoring, diagnostics and control: towards the asset-aware and self-recovery factory. In: 9th IEEE International Conference on Industrial Informatics (INDIN), pp. 497–502, 26–29 July 2011
19. Siddiqui, A.A., Ahmad, A.W., Yang, H.K., Lee, C.: ZigBee based energy efficient outdoor lighting control system. In: 14th International Conference on Advanced Communication Technology (ICACT), pp. 916–919, 19–22 February 2012
20. Serbanati, A., Medaglia, C. M., Biader Ceipidor, U.: Building blocks of the internet of things: state of the art and beyond, deploying RFID—challenges, solutions, and open issues. In: Dr. Cristina Turcu (ed.), 2011, ISBN: 978-953-307-380-4, InTech, Available from: http://www.intechopen.com/books/deployingrfid-challenges-solutions-and-open-issues/building-blocks-of-the-internet-of-things-state-of-the-art-and-beyond
21. Tanaka, K., Higaki, H., Takizawa, M.: Object-based checkpoints in distributed systems. In: Proceedings of the Third International Workshop on Object-Oriented Real-Time Dependable Systems, pp. 9–16, 5–7 February 1997
22. U.S. Department of Energy, Energy efficiency program: test procedure for lighting systems (luminaires). Federal Register 76(150):47178–47180 (2011)
23. Wang, J., Zhang, Y., Lu, G.: Application of WSN in mine emergency communication system. In: 4th International Conference on Wireless Communications, Networking and Mobile Computing, 2008. WiCOM '08, pp. 1–3, 12–14 October 2008
24. Wieland, M., Leymann, F., Schfer, M., Lucke, D., Constantinescu C., Westkämper, E.: Using context-aware workflows for failure management in a smart factory. In: Proceedings of Fourth International Conference on Mobile Ubiquitous Computing, Systems, Services and Technologies UBICOMM 2010, pp. 379–384, Florence, Italy, October, 2010
25. Williams A., Atkinson B., Garbesi K., Rubinstein F.: A Meta-Analysis of Energy Savings from Lighting Controls in Commercial Buildings, Energy Analysis Department, Lawrence Berkley National Laboratory. LBNL PaperLBNL-5095E. Available on-line at: http://efficiency.lbl.gov/drupal.files/ees/Lighting%20Controls%20in%20Commercial%20Buildings_LBNL-5095-E.pdf. Cited 15 January 2012
26. Yoon, J.-S., Shin, S-J., Suh, S.-H.: A conceptual framework for the ubiquitous factory. Int. J. Prod. Res. **50**(8) (2012)

A Service-Oriented Discovery Framework for Cooperating Smart Objects

Marco Lackovic and Paolo Trunfio

Abstract The chapter presents a service-oriented framework designed to support indexing, discovery and selection of network-enabled Smart objects. The framework allows the dynamical discovery of distributed Smart objects and, specifically, the services and operations they provide. To this end, a new metadata model has been defined to describe features, services, and operations of network-enabled Smart objects, and a service-oriented service, accessible through a REST interface, has been implemented for registering, searching and selecting Smart objects on the basis of application needs. The chapter describes the metadata model, the framework architecture and implementation, and the programming APIs.

1 Introduction

Smart objects are items equipped with sensors or actuators, a microprocessor, a communication device, and a power source [13]. Their sensors can provide real-time data such as temperature, pressure, vibrations, and energy measurement. Their communication capability can be limited, such as with RFID tags, or bidirectional such as with IEEE 802.15.4, Bluetooth or low-power Wi-Fi. The Smart object technology is often found with other names such as the *Internet of Things*, the *web of objects*, the *web of things*, and *cooperating objects* but, with the exception of slight differences in the connotations and definitions, they basically represent the same fundamental type of technology. Smart Objects enable a wide range of applications in areas

M. Lackovic (✉)
University of Calabria, Rende, Italy
e-mail: mlackovic@dimes.unical.it

P. Trunfio
DIMES, University of Calabria, Rende (CS), Italy
e-mail: trunfio@dimes.unical.it

G. Fortino and P. Trunfio (eds.), *Internet of Things Based on Smart Objects*,
Internet of Things, DOI: 10.1007/978-3-319-00491-4_5,

such as home automation, building automation, factory monitoring, structural health management systems, smart grid and energy management, and transportation [2, 3].

The goal of our work was to design a service-oriented framework to support the effective indexing, discovery and selection of Smart objects. The large number of proprietary or semi-closed Smart objects systems available today, which has led to partial and non-interoperable solutions, has motivated us to keep a very general approach in designing the framework. We also take in consideration different levels of complexity of Smart objects, including those embedded in everyday objects, such as thermometers, car engines, light switches, and industry machinery, but also more complex devices such as smartphones and personal computers.

The only assumption made in our design is that Smart objects are *IP-reachable*, i.e. they are all connected to an IP-based network and can be reached through their IP address. Despite the reputation of IP as being heavy to implement, it is in fact entirely possible to have lightweight stacks in a Smart object: test networks were described, where this was successfully done [13]. In recent years, to promote the use of IP as the open and interoperable standard for Smart objects, a new open, non-profit worldwide alliance of companies and organizations called the *IP for Smart Objects alliance* (IPSO)[1] has been formed.

The framework proposed in this chapter provides mechanisms for Smart objects indexing, discovery and dynamic selection based on their functional characteristics (the provided services) and non-functional features (the quality of service). *Indexing* means a document containing the description of a Smart object is registered, or published, in a registry, when it is possible in a proactively way (the devices register themselves), to facilitate its fast and accurate finding after a search query. *Discovery* means searching for the Smart objects whose description matches a specified query. *Selection* means finding, among the discovered Smart objects, those that are closer to preference requisites (for example those with higher computing power, with higher battery life, etc.). The term *dynamic* refers to the need to manage the dynamic environment with Smart objects that are often turned on and off, and whose characteristics change over time [4].

The remainder of the chapter is structured as follows. Section 2 discusses related work. Section 3 describes the system architecture. Section 4 describes the metadata model defined to represent the characteristics of Smart objects. Finally, Sect. 5 concludes the chapter by describing the system's APIs, the indexing engine, and the client interfaces used to interact with the system.

[1] IPSO Alliance: http://www.ipso-alliance.org.

2 Related Work

A few research efforts can be found in literature about discovery services for Smart objects. Systems such as *Jini*[2] and *UPnP*[3] provide mechanisms for discovering and using arbitrary network services provided by distributed devices.

Jini is a network architecture for the construction of distributed systems in the form of modular co-operating services. Originally introduced by Sun in 1998, its responsibility has been transferred to Apache under the project name *River*. In a process called *discovery*, a Jini service finds the lookup service (LUS). Once the LUS is found, the *join* process takes place, that is the service register itself providing information such as its id, the object which implements it and other attributes of the service. When a client wishes to make use of a service, it will have first to find the LUS, consult the lookup catalog on the LUS and search based on the type, name or description of the wanted service, and then connect directly to the service through a Java proxy returned by the LUS.

Universal Plug and Play (UPnP) is a set of networking protocols, published in 2008—10 years after Jini, based on established standards such as the TCP/IP, HTTP, XML, and SOAP, that allow networked devices to discover each other's presence on the network and establish functional network services. A key component in a UPnP network is the *control point* which is considered the client in a UPnP network and control devices and services using UPnP protocols. A UPnP compatible device can dynamically join a network, obtain an IP address, announce its name, provide its capabilities upon request, and learn about the presence and capabilities of other devices.

Once a device has established an IP address, the next step in UPnP networking is *discovery*, which is accomplished using a protocol known as the *Simple Service Discovery Protocol* (SSDP). When a device is added to the network, SSDP allows that device to advertise its services to control points on the network. Similarly, when a control point is added to the network, SSDP allows that control point to search for devices of interest on the network. The fundamental exchange in both cases is a discovery message containing a few essential specifics about the device or one of its services, for example, its type, identifier, and a pointer to more detailed information.

Other discovery services get close to the Smart objects world, although they are more related to smart environments than to Smart objects.

Project Aura [5], started in the year 2000, aims to create a system to support computational needs of mobile users by satisfying two competing goals: maximize the use of available resources and minimize user distraction. The system is specifically intended for pervasive computing environments involving wireless communication, wearable or handheld computers, and smart spaces. The system is based on the concept of *personal Aura* which acts as a proxy for the mobile user it represents to marshal the appropriate resources to automatically support the user's task, as he moves from one place to another.

---

[2] Jini, also called Apache River: http://river.apache.org/.

[3] UPnP Forum, UPnP Documents: http://www.upnp.org/resources/documents.asp.

In the Aura architecture, components called *Suppliers* provide the abstract services that tasks are composed of, which are implemented by wrapping existing applications and services to conform to Aura APIs. Such wrappers map the XML-based abstract service descriptions into application-specific settings. When Suppliers are installed in an environment, they become registered with the local *Environment Manager* [11]. Such a registry is the base for matching requests for services: it embodies the gateway to the environment, it is aware of which service suppliers are available to supply which services, and where they can be deployed.

The Voyager framework [10] supports the implementation of dynamically distributed user interfaces, which exploit, on-the-fly, the wireless devices available at a given point in time. Its discovery mechanism relies on proximity considerations to discover ambient devices (i.e., only devices close to the user are considered), and is performed by a component of its architecture, the *Application Manager Interface*. Device discovery is delegated to a single locally running server, which maintains an up-to-date registry of proximate ambient device addresses by performing continuous discovery tests. The client software library receives notifications of newly discovered devices, or devices that go out of range, through a proxy API: this proxy can communicate with the server to maintain a catalogue containing the descriptions of all proximate devices.

All the systems mentioned above may be used with Smart objects but are not specifically designed for their context. A discovery service specifically related to Smart objects has been proposed in [6]: the work describes a framework for building distributed Smart object systems where one of its primary components, called "Smart object wrapper", has a discovery module that allows service advertisement. Smart objects are represented by XML documents that provide meta-information regarding the Smart objects, contain links to the binaries of its services and are also used by the secondary infrastructure *FedNet* to discover the services of the Smart objects and to associate Smart objects with applications.

Even though the last system implements some of the functionalities provided by our framework, it lacks a central indexing registry which makes difficult to use it to register information about a set of Smart objects related to each other for a given purpose. In addition, while the system above uses XML to represent Smart object metadata, in our system we use JSON that offers a more compact metadata representation which in turns results in a more efficient bandwidth utilization. Finally, our system includes a complete API definition and implementation that allow to integrate the discovery service into any existing Smart object middleware.

3 The Architecture

The definition of an architectural model tailored to the application scenarios of reference is a key aspect for the implementation of the framework. Between the options initially considered—centralized and peer-to-peer—a centralized solution has been adopted as it appeared to be the most appropriate in view of the projected

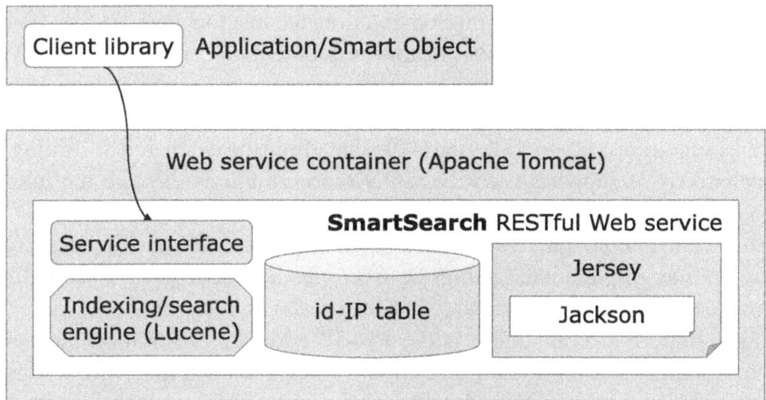

Fig. 1 A schematic representation of the architecture of the system

scenarios. On the other hand, fully decentralized peer-to-peer solutions to resource discovery (as proposed, for instance, in [1, 9], and [12]) could be effectively adopted in larger scale Smart object scenarios.

In the proposed architecture, there is a central registry exposed as a service that indexes the information concerning the Smart objects of a given domain; the search of resources (Smart objects, services or operations) in that domain can be performed by contacting this central registry. Once the resource has been found, a direct reference to it is given so that it may be controlled/queried directly from a client application. To each Smart object it is associated an information provider which has the task to generate the associated metadata.

The centralized architecture model defined, detailed with all its components in Fig. 1, is composed of two main parts:

1. **SmartSearch**: a service that contains a central registry where Smart objects are indexed, and which exposes methods for registering/publishing, searching and selecting Smart objects;
2. **ClientLibrary**: a module, in the form of software library, which allows the application level to interact effectively with the SmartSearch service using objects and local methods, making the remote methods invocation completely transparent.

The SmartSearch service is implemented as a RESTful web service[4] and it is exposed through the Apache Tomcat web service container. REST (Representational State Transfer) is an architectural style for distributed software systems. The term, introduced and defined in 2000 by Roy Fielding (one of the main authors of the HTTP protocol specifications), specifies a series of architectural principles for the design of web services.

[4] RESTful Web Services: http://www.oracle.com/technetwork/articles/javase/index-137171.html.

The framework used in our implementation to develop RESTful web service is Jersey,[5] an open source framework that implements the JAX-RS (Java API for RESTful Web Services) specifications using annotations to map a Java class to a web resource. Specifically, Jersey manages the HTTP requests and the negotiation of the representation, and provides the APIs that allow the extension according to the developer needs. It supports natively JSON representations through the integrated Jackson library.[6]

An important component of the SmartSearch service is the Smart object search and indexing engine, implemented using the free/open source *Apache Lucene* library[7] that provides a high performance and full-featured search engine.

Another important component is the id—IP addresses resolution table, which allows to obtain IP addresses from the Smart objects ids and therefore to reach the Smart object. The presence of such table, along with the fact that the Smart objects are uniquely identified by their id and not by the IP, allows the Smart objects to be able to change their IP in a way completely transparent to the client.

The communication network used is TCP/IP, because we assume the Smart objects have their own IP address. The Smart objects level in fact provides an IP-based protocol for interacting with the Smart objects, as well as the framework for their programming. The communication among Smart objects can be based on IP (according to the IPSO alliance).

The objects and methods of the ClientLibrary component and the methods of the SmartSearch service interface are detailed later in the chapter, in Sect. 5.

4 The Metadata

In order to represent functional and non functional characteristics of Smart objects in a structured way, we defined a *metadata model*. The adoption of metadata models in distributed computing systems is fundamental to manage the heterogeneity of resources and to effectively use them [7, 8]. The proposed model identifies a set of metadata categories useful to index a Smart object in the domain of interest, general enough to satisfy most of the application contexts. The metadata represent the Smart objects static values, while the dynamic ones can be obtained by invoking the appropriate operations on their services.

4.1 Metadata Model

Our metadata model is divided into four main categories:

[5] Jersey: http://jersey.java.net/.

[6] Jackson: http://jackson.codehaus.org/.

[7] Lucene: http://lucene.apache.org.

- **Type**: represents the type of Smart object (e.g., a smart table, a smart wall). Each Smart object will have a tag taken from an enumeration of known tags (a taxonomy) through which it will be identified. This category also contains the id of the Smart object, which allows its unique identification within the system. In case of structured Smart objects, it will also contain the id of the Smart object within which it is contained, and the list of identifiers of the Smart objects contained inside it.
- **Device**: defines the physical characteristics of the device that specify, for example, if the device is running an operating system, what is its name, its version, how many users can use the Smart object, how many objects it can contain, etc.
- **Services**: contains the list of services provided by the Smart object. Each service is provided with an id through which it is possible to interact directly with it, without necessarily knowing in advance the Smart object that exposes it. Every service has a name, a description, the type (sensing or implementation), the return type (boolean, natural number, real number, etc.). A service can be discovered for example by knowing the name or a keyword contained in its description. Each service can also be associated with a quality of service indicator, such as the resolution of the operation performed by the service, its maximum and minimum returned value, and so on. Each service may contain a list of one or more operations that the service can perform. For example, given a "light" service, some operations that may be invoked on the service are "switch on", "switch off" or "invert state";
  - **Operations**: sub-category of Services, it defines the individual operations that may be invoked on a service. As with the Smart object, and for its services, each operation has an id with which the operation can be can reached and invoked, without necessarily knowing the Smart object or service for which it is defined. Each operation is equipped with a series of parameters necessary for its invocation, and a description;
- **Location**: represents the position of the device, which can be indicated in absolute terms, specifying the latitude and longitude, or in relative terms through the use of tags, such as "building B", "floor 3", "room 15". If the position of the Smart object changes over time, then it should be obtained as a service, if it is provided by the Smart object.

The generation of a metadata description document for a simple Smart object can be done by the Smart object creator/manager who, knowing the Smart object in details, can describe its characteristics following the required formalism. For complex Smart objects, this procedure could be accomplished by a module installed on the device, called *information provider*, which has the task to generate the metadata.

In the following tables are listed, for each category, the names of the metadata, their types, a description and their admissible values. It is important to highlight that the metadata represent static values of a Smart object; the dynamic values can be obtained by invoking the appropriate operations on its services.

Geographical information about the location of the Smart object (eg. country, region, province, city, etc.) can be derived starting from the coordinates of the Smart

Table 1 Type metadata

Name	Type	Description	Admissible values
Id	String	It uniquely identifies the Smart object among all the Smart objects known by the system; it may be defined by the user (prior uniqueness check) or automatically generated by the system	Alphanumeric characters, hyphen and underscore, except spaces, commas, quotes, slash, backslash, pipe, etc. (typically the type of the Smart object, followed by a progressive number)
Type	Enumeration	The type of the Smart object	{Table, chair, sofa, desk, floor, ceiling, room, wall, closet, lamp, printer, computer, phone, glasses, camera, display, mirror, speaker, etc.}
So-parent	String	It contains the id of the Smart objects which contains the present Smart object (a smart room contains a smart chair)	Any
So-children	Array of strings	It contains the id of the Smart objects which are contained by the present Smart object (a smart chair is contained by a smart room)	Any

object through different public geographic information repositories. For example, Google Earth and Microsoft Virtual Earth offer some sort of API for geocoding: the use of these resources, however, is governed by precise rules and restrictions. GeoNames[8] instead is licensed under Creative Commons: anyone is free to use as they wish, provided the source is acknowledged: it contains 10 million geographical names categorized into nine classes, in turn sub-categorized for a total of 645 sub-categories (river, lake, park, road, rail, etc.). The data are accessible free of charge through a variety of web services and daily exports of the database.

4.2 Metadata Representation

For the metadata exchange, the JSON (JavaScript Object Notation) text format is used because it is lightweight and easy to read and write manually, as well as to analyze and generate in an automatic way. JSON considerably reduces network traffic compared to XML, which is very important for wireless applications, and is suitable to describe attribute-value pairs.

An example of metadata in JSON format, regarding the description of a smart chair, is shown in Fig. 2. The JSON consists of four members, each corresponding

[8] GeoNames: http://www.geonames.org/.

```
{
  "type": {
    "id": "chair1",
    "type": "chair",
    "so-parent": "office1"
  },
  "device": {
    "users": 1,
    "capacity": 1
  },
  "services": [
    {
      "id": "user-presence1",
      "name": "user-presence",
      "type": "sensing",
      "return-type": "boolean",
      "description": "TRUE: User detected nearby; FALSE: User not detected
nearby",
      "operations": [
        {
          "id": "user-presence-op"
        }
      ]
    },
    {
      "id": "orientation",
      "name": "orientation",
      "type": "sensing",
      "return-type": "real",
      "description": "Orientation from the north direction, clockwise",
      "operations": [
        {
          "id": "orientation-op"
        }
      ]
    },
    {
      "id": "usage",
      "name": "usage",
      "type": "sensing",
      "return-type": "boolean",
      "description": "TRUE: A user is using the device; FALSE: Nobody is using
the device",
      "operations": [
        {
          "id": "usage-op"
        }
      ]
    }
  ],
  "location": {
    "so-nearby": "desk1"
  }
}
```


Fig. 2 JSON representation of a smart chair

```
{
  "type": {
    "id": "smartlamp1",
    "type": "smartlamp",
    "so-parent": "office1"
  },
  "services": [
    {
      "id": "luminosity",
      "name": "luminosity",
      "type": "sensing",
      "return-type": "real",
      "minrange": "0.1",
      "maxrange": "40000",
      "description": "Returns the luminosity of the surrounding environment",
      "operations": [
        {
          "id": "luminosity1"
        }
      ]
    },
    {
      "id": "light",
      "name": "light",
      "type": "acting",
      "minrange": "0",
      "maxrange": "20",
      "description": "Brighten or dims the light",
      "operations": [
        {
          "id": "brighten",
          "params": "{ \"intensity\": \"integer\" }"
        },
        {
          "id": "dim",
          "params": "{ \"intensity\": \"integer\" }"
        }
      ]
    }
  ]
}
```


Fig. 3 JSON representation of a smart lamp

to one of the four metadata categories described above (Type, Device, Service and Location). By looking at the JSON in the example, one can easily derive that the smart chair is located within a smart office (whose id is indicated in the *so-parent* metadata) and offers three sensing services: the presence of a person in its immediate vicinity, the orientation in degrees relative to the north direction, the presence of a person sitting on the chair.

Another example of metadata in JSON format, which is the description of a smart lamp, is shown in Fig. 3.

This JSON has two members Type and Services. As it can be easily inferred by reading the JSON, the smart lamp is included in another Smart object (a smart office denoted by the id *office1* specified in the *so-parent* metadata) and provides one

Table 2 Device metadata

Name	Type	Description	Admissible values
Multitasking	Boolean	Able to execute more operations at the same time	{True, false}
Os-name	String	Operating system name	Any
Os-version	String	Operating system version	Any
Users-number	Integer	Represents the possibility to be used by no one, one or more persons at the same time (e.g. a chair can be used only by one person at the same time, while a room by more persons). −1 represents an unlimited value, not measurable	-1, $\mathbb{N}$
Capacity	Integer	Specifies how many items can be included inside the object	$\mathbb{N}$

sensing service to check the luminosity of the surrounding environment expressed in lux, ranging from 0.1 to 40,000. Two operations are provided to change the light brightness: *brighten* that increases the light brightness by the amount specified as parameter, and *dim* that decreases the light brightness by the specified amount.

5 The Implementation

The methods that are part of the middleware API, which must be invoked at various levels of the architecture to perform the operations mentioned earlier, are listed and described below.

5.1 Service API

The methods exposed by SmartSearch offer the possibility to register a Smart object (the *register* method), search a Smart object (*discover*), obtain the IP address of a Smart object (*resolve*), update the IP address of a Smart object (*updateIP*), remove a Smart object from the registry (*remove*), and get the metadata of a Smart object given its IP (*getJSON*).

A detailed description of such methods follows:

Table 3 Services metadata

Name	Type	Description	Admissible values
Id	String	It uniquely identifies the Smart object service among all the services of all the Smart objects know by the system; it can be defined by the user (prior uniqueness check) or automatically generated by the system	Alphanumeric characters, dash and underscore, except spaces, commas, quotes, slash, backslash, pipe, etc. (typically the name of the service, followed by a progressive number)
Name	String	Name of the service	{Acceleration, magnetic, orientation, light, noise, proximity, gravity, rotation, temperature, humidity, usage, user-presence, wind-speed, wind-direction, gps-coordinates, relative-location, battery-life, plugged, tasks-queue, storage, free-memory, etc.}
Description	String	Textual description of the service	Any
Type	String	Type of service	{Sensing, actuating}
Return-type	String	Returned type	{Boolean, natural, integer, real, string, multi-value}
Power	Real	Consumption in milliamps	$\mathbb{R}+$
Resolution	Real	Accuracy with which the service is able to provide the information	$\mathbb{R}+$
Maxrange	Real	The maximum value/interval provided by the service	$\mathbb{R}+$
Minrange	Reale	The minimum value/interval provided by the service	$\mathbb{R}+$
Operations	Compound	Operations which can be performed by the service, that is the service invocations with specific parameters	List of the operations with the specification of each of the parameters

- **String discover(String query)**: returns the list of Smart objects metadata that match the received query in Lucene format;
- **String resolve(String id)**: returns the IP address of the Smart object, service or operation corresponding to the specified id;

Table 4 Operations metadata

Name	Type	Description	Admissible values
Id	String	Identifies the service operation uniquely among all the operations of all the services of all the Smart objects known to the system; it can be defined by the user (prior uniqueness check) or generated automatically by the system	Alphanumeric characters, dash and underscore, except spaces, commas, quotes, slash, backslash, pipe, etc. (typically the name of the operation, followed by a progressive number)
Params	String	It contains the list of parameters which must be passed to the operation in JSON format	Any
Description	String	Textual description of the operation	Any

Table 5 Location metadata

Name	Type	Description	Admissible values
Latitude	Real	Latitude in decimal degrees (DD)	$\mathbb{R}$
Longitude	Real	Longitude in decimal degrees (DD)	$\mathbb{R}$
Place	String	Textual description of the place in which it is located (eg. the name of the building)	Any
Floor	Integer	Floor of the building in which it is located	$\mathbb{N}$
Room	String	Name of the room in which it is located	Any
So-nearby	Array of strings	It contains the id of the Smart objects permanently close to the current Smart object	Any

- **String register(String metadata, String ip)**: saves in the registry the Smart object metadata, its IP address, and the date and time at which this method was invoked;
- **String updateIP(String id, String ip)**: updates the IP address of the Smart object having the specified id, with the specified IP;
- **String remove(String id)**: removes from the registry the Smart object having the specified id;

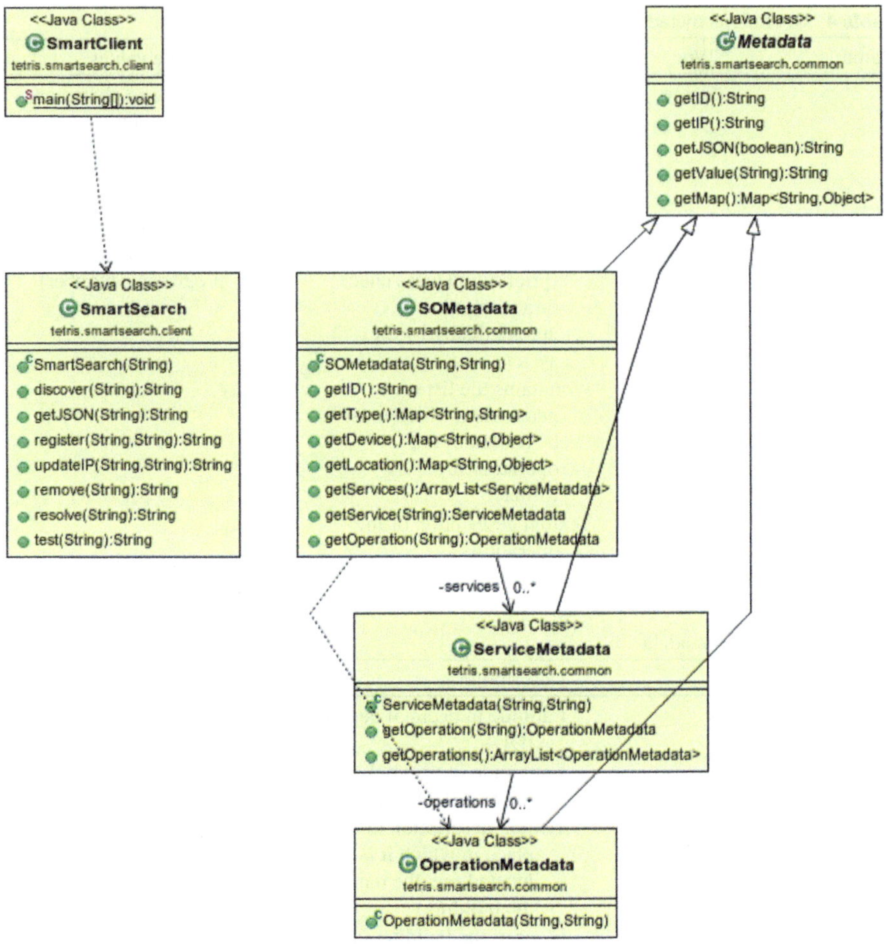

Fig. 4 UML diagram of the classes of the client library

- **String getJSON(String id)**: returns the metadata in JSON format of the Smart object having the specified id.

The IP address, not being a structural information of the Smart object, is passed as a parameter in its own right.

5.2 Client API

The objects and methods provided by the client library are summarized in Fig. 4 through a UML diagram of the classes.

A detailed description of the classes follows:

- **SmartClient**: simple command line application that interacts with the SmartSearch service.
- **SmartSearch**: reference class for the integration with the SmartSearch service. It exposes the methods that allow to invoke the methods of the SmartSearch service.
- **Metadata**: the abstract class containing the elements shared by the classes that model Smart objects, services and operations metadata, and provides the following methods:
  - **String getId()**: returns the id of the element (Smart object, service or operation) to which the metadata applies;
  - **String getIP()**: returns the IP address of the element (Smart object, service or operation) to which the metadata applies;
  - **String getJSON(boolean indented)**: returns the metadata in JSON format. If the specified parameter is *true* then the returned JSON is indented, otherwise not;
  - **String getValue(String key)**: returns the value of the specified attribute;
  - **String getMap()**: returns the metadata of the Smart object as a map, where the key is a string and the associated value can be an ArrayList, a Map, a String or a wrapper object of the basic types (Integer, Boolean, etc.).
- **OperationMetadata**: contains the metadata of an operation provided by the service of a Smart object.
- **ServiceMetadata**: contains the metadata of a service of a Smart object and provides the following methods:
  - **OperationMetadata getOperation(String id)**: returns the OperationMetadata object corresponding to the operation having the specified id. If such operation is not provided by the service then *null* is returned;
  - **ArrayList<OperationMetadata> getOperations()**: returns the list of operations provided by the service.
- **SOMetadata**: contains the metadata of a Smart object and provides the following methods:
  - **String getID()**: returns the id of the Smart object on which the method is invoked;
  - **Map<String, String> getType()**: returns the map containing the metadata of the group *Type*;
  - **Map<String, Object> getDevice()**: returns the map containing the metadata of the group *Device*;
  - **Map<String, Object> getLocation()**: returns the map containing the metadata of the group *Location*;
  - **ArrayList<ServiceMetadata> getServices()**: returns the map containing the metadata of the group *Services*;
  - **ServiceMetadata getService(String id)**: returns the ServiceMetadata object of the service of the Smart object having the specified id;

- **OperationMetadata getOperation(String id)**: returns the OperationMetadata object of the operation having the specified id.

The language used for the queries is the one supported by *Apache Lucene*.

5.3 Indexing

The SmartSearch service has been implemented as a *RESTful Web service* using *Jersey*,[9] an open source framework that implements JAX-RS (Java API for RESTful Web Services) using annotations to map a Java class to a Web resource, and natively supports JSON representations through the integrated library *Jackson*.[10]

The service discovery prototype is written in Java (version 1.7) and has been deployed on *Apache Tomcat*[11] (version 7.0) installed on a server machine having Ubuntu Server[12] (version 12.04) as operating system, but may be installed on any Linux distribution. There are no special hardware requirements, given the relatively small amount of memory required by Lucene. Regarding the ClientLibrary, it can be used on any computer with a Java Development Kit (version 1.7 or higher), on any operating system with no particular requirements in terms of hardware, except those imposed by the application that uses the library.

5.3.1 Indexing Engine

The core component of the SmartSearch service is the indexing and search engine; this component has been implemented by using the open source library *Apache Lucene* (version 4.2.0), an open source library that implements a performant search engine and rich of functionalities.

Another important component is the *id-IP Table*, a directory service that maps Smart objects, services and operations to IP addresses. This table, along with the fact that Smart objects are uniquely identified by their ids and not by their IP addresses, allows for the Smart objects to change their IP addresses in a transparent way to the client.

The Apache Lucene library provides a scalable and high-performance indexing system with the following characteristics:

- low memory requirements (only 1MB of heap);
- incremental indexing as fast as the batch one;
- size of the index about 20–30 % of the indexed items.

Lucene also includes efficient search algorithms that provide:

---

[9] Jersey: http://jersey.java.net/.

[10] Jackson: http://jackson.codehaus.org/.

[11] Apache Tomcat: http://tomcat.apache.org/.

[12] Ubuntu: http://www.ubuntu.com/server.

- classified search—the best results are returned first;
- a rich query language that supports a variety of functionalities, including:
  - single and multiple wildcard searches;
  - fuzzy searches based on sophisticated algorithms such as the Levenshtein distance (a measure of the difference between two strings): for example, a fuzzy search of the word *roam* will find terms like *foam* and *roams*;
  - search by range that allow to identify objects whose fields have the values within the specified ranges;
  - boolean operators allow terms to be combined through logic operators.
- search on date ranges;
- sort on any field;
- allow updating and searching simultaneously.

5.3.2 Indexing Structure

The data managed by Lucene are represented as *documents* with text *fields* that contain classified information about the documents and allow to perform searches on them. Hence, each Smart object metadata is saved as a Lucene document, whose fields are the individual metadata of the Smart object.

The Lucene document corresponding to a Smart object consists of two fields: (1) an "id" field containing the id of a Smart object, and (2) a "json" field containing the metadata of the Smart object in JSON format.

5.4 Client Interfaces

To demonstrate the functionality of the service two types of clients have been developed: the first one uses JSP web pages published by a web server which can be accessed from a browser; the second one is a command line interface that can be used from terminal. Sections 5.1 and 5.2 describe such components.

5.4.1 JSP Client Interface

A Web-based user interface has been implemented to expose all the discovery service methods discussed above. A screenshot of such interface is shown in Fig. 5.

The JSP interface consists of a set of web pages that allow to interact with the SmartSearch service through the invocation of its methods. Specifically it consists of a page, shown in Fig. 5, which displays the list of all the Smart object indexed: for each one of them is shown its type, its IP address, the list of the services provided by it, and the date on which it was registered in the index. By clicking on a Smart object or service id on that page, the respective metadata will be shown and by clicking

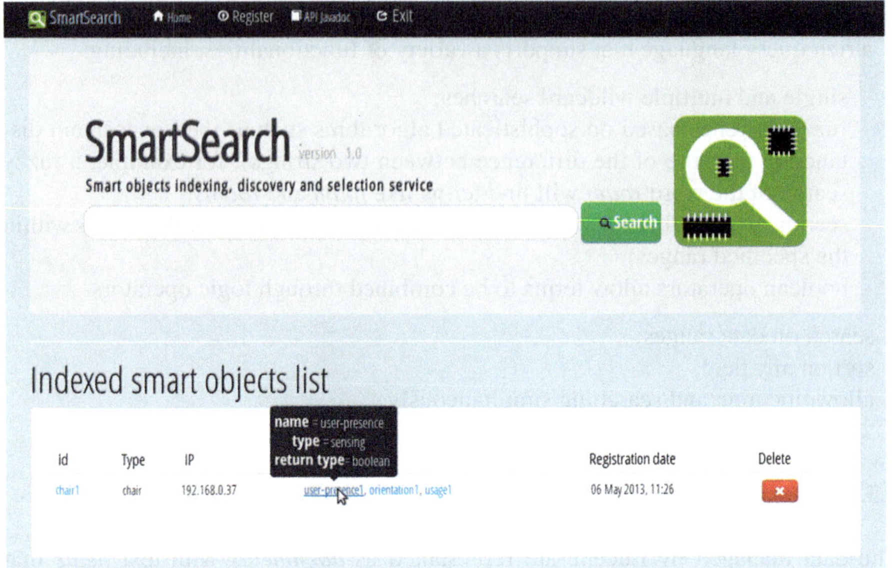

Fig. 5 JSP client interface

on the red button in the *Delete* column the respective Smart object will be removed from the index.

Another page is dedicated to the registering of a Smart object: the registration is done simply by entering the IP address of the smart obejct in the proper field, choosing the file which contains the metadata in JSON format of the Smart object to be registered and by pressing a button to submit the request. If successful, a message will be displayed to confirm the registration of the Smart object, otherwise an error message will be displayed. Other pages show the Javadoc of the service and the client library APIs.

5.4.2 Terminal Client Interface

The command line application that interacts with the SmartSearch service consists of a jar file, called SmartClient, and can be executed by running the following command:

```
java -jar smartClient.jar address command parameter1 [parameter2]
```

where *address* is the server hostname or IP address where the service is deployed, while *command* must be one of the following:

- **register** = registers a Smart object in the service registry
  parameter1 = name of the file containing the metadata in JSON format
  parameter2 = IP address of the Smart object
- **discover** = performs a search
  parameter1 = query in Lucene format

```
:~$ java -jar smartClient.jar smartsearch.dimes.unical.it getJSON chair1
Connecting...
Invoking getJSON(chair1) on smartsearch.dimes.unical.it...
Result:
{
  "type" : {
    "id" : "chair1",
    "type" : "chair",
    "so-parent" : "office1"
  },
  "device" : {
    "users" : 1,
    "capacity" : 1
  },
  "services" : [ {
    "id" : "user-presence1",
    "name" : "user-presence",
    "type" : "sensing",
    "return-type" : "boolean",
    "description" : "TRUE <- User detected nearby; FALSE <- User not detected nearby"
  }, {
    "id" : "orientation1",
    "name" : "orientation",
    "type" : "sensing",
    "return-type" : "real",
    "description" : "Orientation from the north direction, clockwise"
  }, {
    "id" : "usage1",
    "name" : "usage",
    "type" : "sensing",
    "return-type" : "boolean",
    "description" : "TRUE <- A user is using the device; FALSE <- Nobody is using the device"
  } ],
  "location" : {
    "so-nearby" : "desk1"
  }
}
```

Fig. 6 A terminal client interface

- **getJSON** = returns the metadata of the specified Smart object
  parameter1 = Smart object id
- **resolve** = returns the IP address of the specified Smart object
  parameter1 = Smart object id
- **updateIP** = updates the IP address of the specified Smart object
  parameter1 = Smart object id
  parameter2 = the IP address of the Smart object
- **remove** = removes the specified Smart object from the service registry
  parameter1 = Smart object id

Figure 6 shows an examples of invocation of the getJSON method from command line.

6 Conclusions

Providing effective indexing and discovery services is fundamental in Internet of Things scenarios, where a large number of heterogeneous Smart objects must be dynamically discovered and integrated with each other to satisfy users' and applications' needs, based on the services and operations they provide. We worked in this direction by designing and implementing a service-oriented framework that supports indexing, discovery, and dynamic selection of Smart Objects.

We have showed how the proposed framework can provide mechanisms for Smart objects indexing, discovery and dynamic selection based on their functional characteristics (the provided services) and non-functional features (the quality of service). We have defined a metadata model useful to represent functional and non functional characteristics of the Smart objects in a structured way; the proposed model identifies a set of metadata categories useful to index a Smart object in the domain of interest, general enough to satisfy most of the application contexts. We have implemented a prototype, describing in details its API.

The comparison with similar systems showed that our framework combines a series of distinctive features: (i) a Smart Object registry accessible as a network service; (ii) the use of JSON for a more compact metadata representation, which results in a better bandwidth utilization; (iii) the use of modern Web standards and technologies (e.g. REST); (iv) an open design, which makes the system suitable for integration in many infrastructures.

Acknowledgments The authors would like to thank Giovanni Cirelli for his implementation efforts. This work has been partially supported by TETRis - TETRA Innovative Open Source Services, funded by the Italian Government (PON 01-00451).

References

1. Cannataro, M., Talia, D., Tradigo, G., Trunfio, P., Veltri, P.: SIGMCC: a system for sharing meta patient records in a peer-to-peer environment. Future Generation Computer Systems **24**(3), 222–234 (March 2008) NULL
2. Dunkels, A., Vasseur, J.: Ip for smart objects. White paper 1, September 2008. Internet Protocol for Smart Objects (IPSO) Alliance
3. Fortino, G., Guerrieri, A., Russo, W., Savaglio, C.: Middlewares for smart objects and smart environments: Overview and comparison. In: Fortino, G., Trunfio, P. (eds.) Internet of Things Based on Smart Objects: Technology, Middleware and Applications, Springer (2014)
4. Fortino, G., Lackovic, M., Russo, W., Trunfio, P.: A discovery service for smart objects over an agent-based middleware. In: Proceedings of the 6th International Conference on Internet and Distributed Computing Systems (IDCS 2013), Lecture Notes in Computer Science, vol 8223, pp. 281–293. Springer, Hangzhou, China, September 2013 ISBN 978-3-642-41427-5
5. Garlan, D., Siewiorek, D.P., Steenkiste, P.: Project aura: toward distraction-free pervasive computing. IEEE Pervasive Comput. **1**, 22–31 (2002)
6. Kawsar, F., Nakajima, T., Park, J.H., Yeo, S.-S.: Design and implementation of a framework for building distributed smart object systems. J. Supercomput. **54**(1), 4–28 (2010)
7. Mastroianni, C., Talia, D., Trunfio, P.: Managing heterogeneous resources in data mining applications on grids using xml-based metadata. In: Proceedings of the 17th International Parallel and Distributed Processing Symposium (IPDPS 2003), Nice, France, April 2003
8. Mastroianni, C., Talia, D., Trunfio, P.: Metadata for managing grid resources in data mining applications. J. Grid Comput. **2**(1), 85–102 (2004)
9. Pirrò, G., Talia, D., Trunfio, P.: A dht-based semantic overlay network for service discovery. Future Gener. Comput. Syst. **28**(4), 689–707 (2012)

10. Savidis, A., Stephanidis, C.: Distributed interface bits: dynamic dialogue composition from ambient computing resources. Pers. Ubiquitous Comput. **9**(3), 142–168 (2005)
11. Sousa, J.a.P., Garlan, D.: Aura: an architectural framework for user mobility in ubiquitous computing environments. In: Proceedings of the IFIP 17th World Computer Congress—TC2 Stream / 3rd IEEE/IFIP Conference on Software Architecture: System Design, Development and Maintenance, WICSA 3, pp. 29–43, Kluwer, Deventer, The Netherlands (2002)
12. Talia, D., Trunfio, P., Zeng, J.: Peer-to-peer models for resource discovery in large-scale grids: a scalable architecture. In: Proceedings of the 7th International Conference on High Performance Computing in Computational Sciences (Vecpar 2006), LNCS, vol. 4395, pp. 66–78. Springer, Rio de Janeiro, Brazil (2007)
13. Vasseur, J.-P., Dunkels, A.: Interconnecting Smart Objects with IP: The Next Internet. Morgan Kaufmann Publishers Inc., San Francisco, CA, USA (2010)

Smart Manufacturing Through Cloud-Based Smart Objects and SWE

Pablo Giménez, Benjamín Molina, Carlos E. Palau, Manuel Esteve and Jaime Calvo

Abstract Smart manufacturing is a key aspect for innovation and competitiveness, and involves several dimensions of the production chain to be analysed, assessed and enhanced within a factory. To target this issue, concepts and ideas behind the IoT (Internet of Things) are applied, so that connected smart entities cooperate in order to achieve broader goals or increase the overall knowledge in the factory through information sharing. Smart entities in the IoT are typically referred as WSNs (Wireless Sensor Networks) that capture physical (real) data and events and produce virtual (digital) information to be processed. Unfortunately, current WSNs have limited interoperability and processing capabilities, reducing the integration degree with existing applications. This chapter proposes a solution for both previous technical challenges within a factory. Interoperability is achieved by means of SWE (Sensor Web Enablement) whereas processing capabilities are provided through virtualizing smart objects in a datacentre, placed commonly in the factory but it could also be located elsewhere, applying cloud-based techniques. The architecture and deployment has been arranged for the specific use case of a manufacturing company and a risk prevention scenario. Experimentation results show that smart objects could be provided at runtime with fine granularity level depending on the tasks to be performed. Moreover, smart objects are able to co-operate forming meta-objects

P. Giménez (✉) · B. Molina · C. E. Palau · M. Esteve
Universitat Politècnica de València,València, Spain
e-mail: pabgisa@upvnet.upv.es

B. Molina
e-mail: benmomo@upvnet.upv.es

C. E. Palau
e-mail: cpalau@upvnet.upv.es

M. Esteve
e-mail: mesteve@upvnet.upv.es

J. Calvo
Universidad de Salamanca,Salamanca, Spain
e-mail: Jaime.calvo@usal.es

G. Fortino and P. Trunfio (eds.), *Internet of Things Based on Smart Objects*,
Internet of Things, DOI: 10.1007/978-3-319-00491-4_6,

to satisfy global tasks or minimize certain risks. Finally, smart objects are able to encapsulate private (health and/or personal) data that should not be shared with other objects or processes.

Keywords Smart objects · Industrial safety · Wireless sensor networks · Sensor observation service

1 Introduction

Security technological advances in industrial environments have evolved considerably in recent years but there are still risks concerning worker's safety and health. Therefore new integrated approaches from different scopes (legal, technological, socio-economic, etc.) should be built to ensure the continuous safety and wellness of workers in the handling, machining and assembly factories, allowing workers to become key factors of competitiveness and differentiation of the new production model. In this chapter, we propose the usage of smart virtualized objects to perform intelligent tasks such as increasing productivity and minimizing risks. The architecture presented will be later deployed on a test bed in order to assess a special risk (collision detection).

The IoT is evolving from simple sensors with network connectivity to a collection of interrelated and interconnected objects, called Smart Objects (SO) [1–5]. Currently there is an undeniable trend of increasing computing capabilities and using emerging technologies such as Near-Field Communications (NFC), real-time localization and embedded sensors of everyday objects to turn them into Smart Objects [6]. These SOs are fully functional on their own, but added value is obtained through communication and distributed reasoning. Such objects have become building blocks for the IoT and enable novel computing applications.

The proposed system in this chapter consists basically of multiple objects and each one may be perceived as a Wireless Sensor Network (WSN) capable of obtaining multiple environmental data. WSNs are a set of small, low cost and low energy devices that monitor an area of interest, and enable applications to obtain up-to-date information about the physical world. The data is stored and processed according to the scenario of application. This information is especially valuable for environments in which it is inefficient, difficult or dangerous for people to collect data on site by themselves, e.g. agriculture, maritime, healthcare, industrial and even military applications.

WSNs are one of the most important elements in the IoT paradigm, as they provide a virtual layer where any piece of sensing capable information about the physical world can be potentially accessed by any computational system. The use of standards such as 6LowPAN, defined by IETF, allows transmitting IPv6 packets through computationally restricted networks [7]. Even if WSN's use and popularity is increasing they are still subject to huge research in several fields. Most of the research has been placed on reducing energy consumption [8], developing efficient routing protocols [9, 10] and introducing or enhancing security issues [11].

Sensors and information gathering from the physical environment are incorporated to everyday life in multiple environments. A way of cheaply incorporating sensors into such environments and testing their suitability is the usage of virtual sensors. A virtual sensor is a software component that emulates a sensor and sends periodically information, inserting observations and updating its position. In order to test the feasibility and scalability of large systems it is necessary to introduce simulation tools, as it would be extremely expensive to deploy hundreds of sensors to replicate any potential risk scenario. Note that a simulated system may evaluate all types of situations as exceptional cases or long-term behaviour whereas a real factory may only require some of the sensor networks targeting those risks that are relevant for them. The usage of simulation tools for testing environments allows a faster, cheaper and riskless deployment timeline [12, 13]. Typically, WSN emulators have focused in the analysis of communications, energy consumption or security. Our proposal focuses on the upper layers on the data generated by the sensors and how this information is made available and consumed by user applications.

Besides simulation tools, there are standards (protocols and specifications) that target interoperability among sensor networks. Currently real sensor networks are vendor specific and custom applications and adaptations are required if different sensor networks from different vendors need to interoperate. Because the SDI (Spatial Data Infrastructures) did not have a generic framework to integrate data from sensors, it was necessary to extend the specifications. For such reason the Open Geospatial Consortium (OGC) founded Sensor Web Enablement to set standards for accessing and controlling sensors and sensor networks over the Internet. The initiative promotes interoperability, defining various services and components. The presented work focuses on the Sensor Observation Service (SOS) which supports a common data registration model from any sensor, and also a common data query and retrieval model from any process entity [14, 15].

The SOS is used as a sink for all measurements provided by sensors, meeting the standards set by the SWE, allowing the inclusion of any type of sensor in the system and sending the measurements easily over an IP network, including the Internet. The only restriction is the capacity of the SOS that in any real or virtual deployment may need to be distributed, in order to support all data flows generated by the sensors.

A preliminary version of this initial design was presented in [16] with some interesting results that will be extended in this chapter. Whereas the preliminary version focussed on a centralized reasoning entity based on a CEP (Complex Event Processing) this chapter covers the same problem statement from a different perspective introducing smart objects, which allow an interesting distributed information management mechanism.

Besides capturing all available data present in a particular scenario (in our case a factory) it is important to process and correlate it in order to detect relevant events even before accidents occur. A typical approach consists in centralizing the processing capability in a single entity that is configured to track special situations. The processing entity may be also deployed in a distributed way so that multiple processes monitor different situations in a modular case. In any case, processing entities are separate from WSNs, which means that there is a part in charge of sensing and another for detecting events. The concept of smart object unifies both concepts and provides a single logical entity able to understand what is happening internally (through its own WSN) and what is happening externally (through communication activity). Understanding here means far more than just sensing and sending messages, but the ability to reason on the obtained data and to provide a higher level of interaction and cooperation with the outside world.

The chapter is structured as follows. Section 2 presents the motivation and related work. In Section 3 the outline of the system is described. Section 4 describes the system and its main functionality. Section 5 evaluates the system in an operational scenario with a particular use case. The chapter ends with the conclusions and further work.

2 Motivation and Previous Work

Smart Objects are emerging entities increasing significantly as new Internet connectivity points, and provide a set of new resources to be consumed by networks, services and applications. The SO paradigm provides the relevant features to embed new capabilities into an everyday object, allowing extended access information up to complex services invocation and interaction. A SO typically provides the following features [17, 18]:

- A unique identification
- Capability to communicate effectively with its environment
- Data storage about itself
- A language to display its features and its needs over its lifecycle
- Capability to participate in or making individual decisions relevant to its own destiny
- Capability for surveying and controlling its environment
- Generation of interaction by services offering: contextual, personal and reactive services

There are already many existing applications using SOs such as INOX [19], a service management platform. It integrates the features of IoT and provides the functionality which allows for a better use of sensors. Another example refers to a system that transforms a product into a RFID SO capable of acting by itself and achieving its objective throughout its life cycle [20].

One of the most innovative aspects proposed in this chapter is how SOs are managed through virtualization. To the extent of our knowledge, there is little approach in this direction [21]. Considering sensors (and even WSNs) as small systems with limited (processing) capabilities, it makes sense to virtualize its capabilities in a data centre, so that sensors (WSNs) perceive that they are as powerful as normal computers. The idea seems logical for indoor environments, where a datacentre may be placed nearby and there is no need for too much communication between the real WSN and the virtualized SO. For outdoor environments, communication requirements may guide to a non-practical or non-viable proposal.

The technological approach is driven by cloud computing technologies [22, 23]. Cloud computing is a model for on-demand access to a shared pool of configurable computing resources (e.g., networks, servers, storage, applications, services, and software) that can be easily provisioned as and when needed. Cloud computing provides an abstract interface, aggregates resources to gain efficient utilization and allows users to scale up to solve larger science problems. It enables the system software to be configured as needed for individual application requirements. We will use cloud computing for creating and managing SOs as needed; some SOs may be available continuously whereas others may be created on the fly for a short (or long) period of time. It mainly depends on the configuration of the system: a worker may have a corresponding SO while it is at the factory, which can be a short time or a long time. A fixed machine continuously working which is always tracked by sensors will have a continuous SO. Other machines working from time to time will only require temporal (on the fly) SOs.

Factory workers (e.g. automotive) typically employ different machinery from various manufacturers that use their own set of sensors to control and automate operation. For a big factory, it seems reasonable to unify the potential huge amount of different WSNs in a centralized environment. For this purpose an interoperability mechanism is necessary. Another innovative aspect in this chapter is the use of OGC SWE in indoor environments, which will be described below.

The Open Geospatial Consortium created the SWE [14] as a group of specifications covering sensors, related data models and services that offer accessibility and control over the Web. The SWE architecture is composed of two main models: the information model and the service model (see Fig. 1). The information model describes the conceptual models and encodings whereas the service model specifies related services.

The conceptual models refer to: transducers, processes, systems and observations. The information model also includes a core suite of language and service interface specifications, such as: (i) Transducer Markup Language (TML), currently deprecated; (ii) Sensor Model Language (SensorML); (iii) Observation and Measurements (O&M).

On the other hand, the service model describes the SWE framework services, which include: (i) Sensor Alert Service (SAS); (ii) Sensor Planning Services (SPS); (iii) Web Notification Service (WNS); (iv) Catalog Service Web (CSW); and (v) Sensor Observation Service (SOS).

The SWE architecture component that has been used and analysed in this document is the SOS [15]. The main purpose of this service consists in allowing access to sensor observations in a standard way for any sensor system, including in-situ, remote, fixed and mobile sensors. The SOS complies with the O&M specification for modelling sensor observations, and with the SensorML specification for modelling sensors and sensor systems.

The used SOS implementation is 52 north [24], which has been developed in Java.

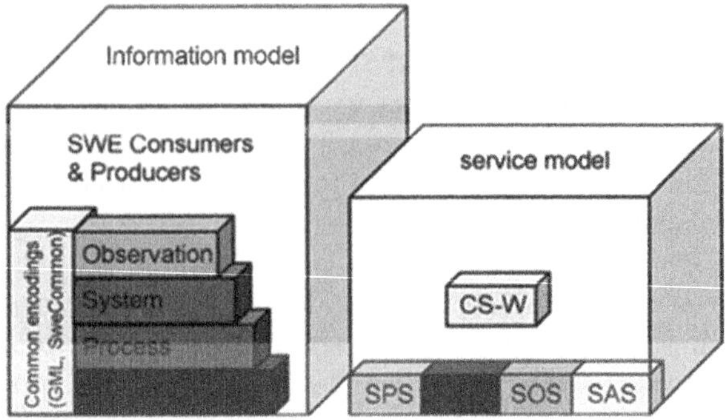

Fig. 1 Main blocks of the SWE architecture [14]

Traditionally, simulators and emulators are useful tools for networking research as they simulate or emulate real networking protocols and provide a controllable environment for studies. This feature is even greater for WSN applications, because real WSNs are in frequent upgrades and their deployment is tightly embedded in the physical environment. As evidence, several sensor network simulators and emulators have been developed for large-scale WSN studies. Use of WSNs has as main goal the detailed knowledge of a certain environment or scenario. It implies sensing a great number of features and conditions that can only be achieved by means of deploying a large amount of WSNs. However, such situation is more theoretical and less practical, and typically users and administrators perform a gradual deployment of WSNs, depending on critical features to sense or available budget. Because of the importance of the simulation, there are already several tools that facilitate the study of sensor networks, such as TOSSIM [25], EmStar [26], NS-3 [27] OMNeT++ [28], and Glomosim [29]. However no previous simulator or emulator is suited to our needs, because none of them have been designed to natively interoperate with SWE architecture, and specifically with the SOS.

3 Outline of the System

The system architecture has two main parts on a high-level basis. On one hand the data gathered from the sensors (WSNs) are inserted in the SOS implementation. One or more WSNs are emulated by means of a Sensor Simulator [16] although real WSNs may even coexist in the environment. There is a Control Center (CC) that aggregates different applications that make decisions based on the information available in the SOS. WSNs, CC and SOS may be connected by wired or wireless communication networks. The HMI is responsible for offering a web monitoring platform, as will be described later. The Event Processing Module is in charge of proactively detecting alarms based on sensed data in the SOS. The Action Handler performs the corresponding actions once an alert has been detected. The use case in Sect. 5 will clarify this issue (See Fig. 2).

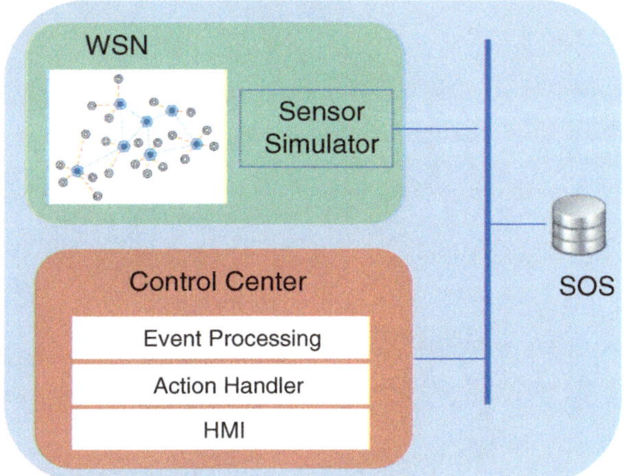

Fig. 2 High-level system architecture. Collection data [16]

In the first version presented in [16], the CEP was the only entity able to generate alarms from SOS data, but in our new proposed architecture SOs are able to generate more specific alarms from WNS and they are also able to sense the environment either internally or externally.

Typical WSNs are only able to measure some data and have limited processing power. In order to increase their capacity we propose the use of SOs to virtualize really powerful WSNs in the system. In this way, each SO is able to obtain all the sensed information needed from a SOS as well as any available resource available in a Common Resource Repository (CRR). With such inputs, the virtualized SO is able to perform not only powerful tasks but also take into account its environment (factory). The processing capacity of each SO has two main objectives:

- To be able to perform powerful tasks in real time (e.g. monitoring)
- To be able to communicate with other SOs to perform cooperative or notification tasks. Though each SO has access to environmental facilities (CRR) and measurements (SOS), it is unaware of things going on with other. Considering a factory with a significant amount of SOs, it seems sensible to provide interaction among SOs. Two ways have been envisioned (see Fig 3):
  - Event notification: each SOS generates a certain event and sends it to the system through a Notification Center (NC) module. Any other SO in the factory can subscribe to this event in order to receive notification whenever it is triggered.
  - Multiple co-operation: SOs may cooperate in order to achieve a common goal through consensus. From this point of view they compound a logical meta-SO. Figure 3 depicts an example of meta-SO called Scheduler: here some SOs may decide, plan and schedule together a certain activity. The test case that will be presented later takes into consideration collision avoidance through a common schedule of mobile machines.

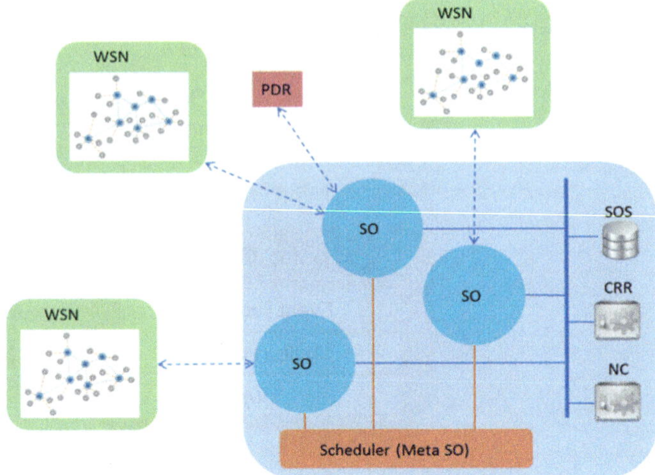

Fig. 3 High-level system architecture. Smart objects

4 Design and Development

The system is designed to generate and virtualize SOs on demand as they are needed. Cloud computing strategies are used in order to address this goal. There are some initial objects already available in the system (a factory in the scenario under study) providing basic support to deployed WSNs. As example and test case we will use collision detection between automated machinery which will facilitate the comparison with our previous work [16]. Using smart objects that monitor in real time their environment and generate notification through events allows a powerful and decentralized risk management system.

There is a need to simulate the sensed data by means of the developed OGC sensor simulator. It is a software component that allows the simulation of multiple sensors of different nature, in order to represent all possible scenarios and use cases, avoiding the expensive deployment of real sensors. Additionally the simulator can be used only with virtual sensors or in conjunction with real sensors. The goal of the simulator is twofold: (i) testing applications that require OGC-based sensor data and (ii) evaluate the feasibility of different deployments prior to do them with real equipment.

4.1 Description of the System

Though the system may be described from a general perspective, we will particularize the description from the specific use case point of view, in order to facilitate the understanding; focusing the research on a special process within virtualized objects and detect special aspects of the use case to consider. Each WSN will have its own smart object that allows it to manage and process all the information. Besides, each WSN will behave as normally and will regularly send measurements (e.g. position, speed and other measurements) to the SOS. Thus, the WSN has two interfaces: (i) a real interface to the SOS to send measurements, and (ii) a virtual interface to a virtual internal process unit modelled by a SO and thus extending its internal processing power. In our test case related to collision prevention between lift trucks in a factory the real interface will send (at least) position and speed measurements to the SOS and the virtual interface will monitor in real time if there is potential collision risk.

There are several ways to calculate or estimate potential risk for a SO. First, the SO must know the position of all possible crosses involving some risk. For this purpose the SO queries the Common Resource Repository (CRR) for a factory map including all intersections. Then the SO obtains from the SOS position and speed of the lift trucks. Note that this information may also be obtained directly from the WSN (in fact it would be more efficient) but we develop a facility to obtain data (position and speed) from the SOS of any truck; note that if a SO would need to retrieve such information from a different WSN it would need to directly contact the SOS.

From such information (position, speed and map) the SO can estimate in real time whether it is entering a risk area which may typically refer to an intersection (see [16]). In case of collision risk, the SO notifies the event to the notification center (NC). If another lift trucks enters the risk area of the same intersection, the correspondent SO will receive the notification through the NC that serve as communication channel among SOs. Note that several events may be defined associated to the same risk or different actions might be taken when the event is notified (e.g. slow down, stop, etc.) depending on internal or environmental status.

As commented previously, besides the event notification system there is also a cooperative interaction among SOs. In order to avoid collisions before they happen, SOs may co-operate forming a meta-object called scheduler. The scheduler allows getting the optimal initial route for each SO before starting the drive. In this way one can avoid collisions before they occur. It has to be highlighted that the scheduler is created on real time by the association of SOs. This meta-object may disappear once the initial route (for each SO) has been estimated or it may remain to monitor, track and readjust if necessary for a certain while.

Another relevant feature of SOs consists in the encapsulation of personal and private data (e.g. medical data). The SO is able to cross-check personal data with environmental data and extract conclusions (e.g. a worker associated to a WSN might not be exposed to specific amines longer than a certain amount of time) without disclosing sensitive information. The SO is the only entity in the system that has access to Personal Data Records (PDRs) or medical records and therefore knows the risks of a specific person/worker. It is completely private and confidential, because the only one who reads the information is the SO and then gets the necessary information from the SOS or CRR to perform some reasoning process.

4.2 Insertion of Data in the SOS

The simulator replaces an individual or a group of sensors, and the key communication component at application level is the interaction with the SOS. The simulator not only generates data with a certain pattern, for each individual sensor, but inserts their observations in the SOS. The main advantage is that the interface is provided via web (HTTP) so that any emulated sensor can easily communicate with the SOS (i.e. registering and inserting observations). The SOS is a service that basically offers two levels of interface:

- An interface for sensors: the process consists in registering each sensor in the SOS and sending measurements. The first step is performed by means of a *registerSensor* operation, which allows saving a new sensor. Once the sensor has been registered, it can start sending measurements at certain intervals, which depends either on the physical quantity being measured or the need of control required. The operation is called *insertObservation*. The SOS supports both fixed and mobile sensors. In the latter case, mobile sensors must send (besides measurements in the *insertObservation* operation) their current location. The operation is called *updateSensor*.
- An interface for external processes, through which any application can access historical data (even real-time data) regarding any registered sensor. Note that, as the SOS service centralizes all sensors, it is possible to search and apply simple spatio-temporal filters, e.g. "get all sensors that monitor temperature" or "get all sensors located in an area".

Note that there are currently two versions of OGC SWE, v1.0 and v2.0. The diagram in Fig. 4 refers to version 1. The new version (2.0) has several enhancements, but the main functionalities remain as in the previous version. For example, the *RegisterSensor* operation has been replaced by the *insertSensor* operation in OGC 2.0. Authors of this chapter are currently porting the architecture from OGC 1.X to 2.0.

4.3 Functionality of the Simulator

The simulator is configured through an XML file that includes a list of all sensors to be simulated with their description and specific configuration. Depending on the scenario and the use case, different sensors may be used. Some examples of sensors to be simulated are mobile sensors, such us workers in a factory (some property of the worker, or just its location), temperature sensors, humidity sensors, sound sensors, chemical sensors, machines (some property of the machine, or just its location), etc. For each sensor the following parameters may be specified:

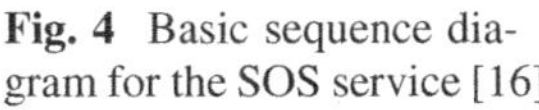

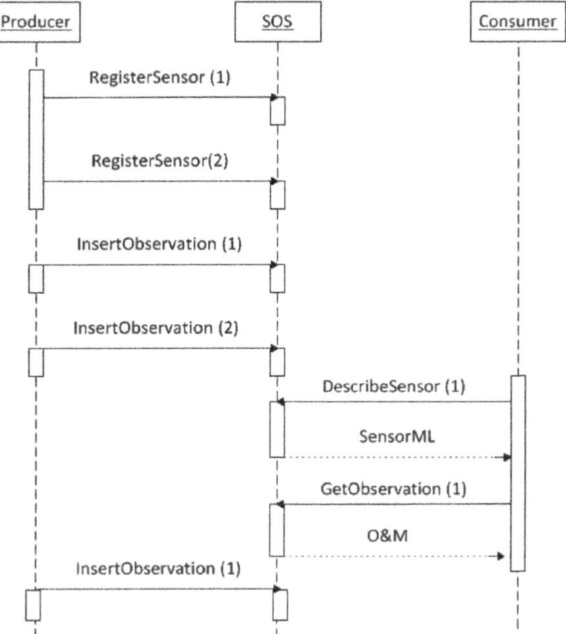

Fig. 4 Basic sequence diagram for the SOS service [16]

- Minimum and maximum provided by each sensor.
- Periodicity model (how often data is sent to the SOS). This is extremely important as it is influenced by data loss (some packets will not be sent to the SOS) and delay (some packets will not be sent to the SOS in time) within a real WSN.
- Simulation mode (sine, Gaussian, exponential, etc.). This mode describes how data is generated during the simulation. Additionally, each mode is able to simulate measurement errors and anomalies. Observations may be retrieved from a stored file including real data.
- Location (latitude and longitude). This property can be random or linear; the linear mode is useful for simulating moving paths and collisions. Though there may be advanced mobility models, we are currently interested in linear ones, simulating people and vehicles.

For simplicity and interactivity, the simulator includes a graphical user interface which displays all sensors to simulate. Users can stop each sensor separately (independently) to track in detail a single sensor. It also allows loading a new configuration file or adding new sensors.

4.4 Simulator HMI

The used simulator is capable of generating measurements from a high number of sensors considering also measurement errors and anomalies. The GUI is depicted in Fig. 5. It basically displays all current information regarding any created sensor.

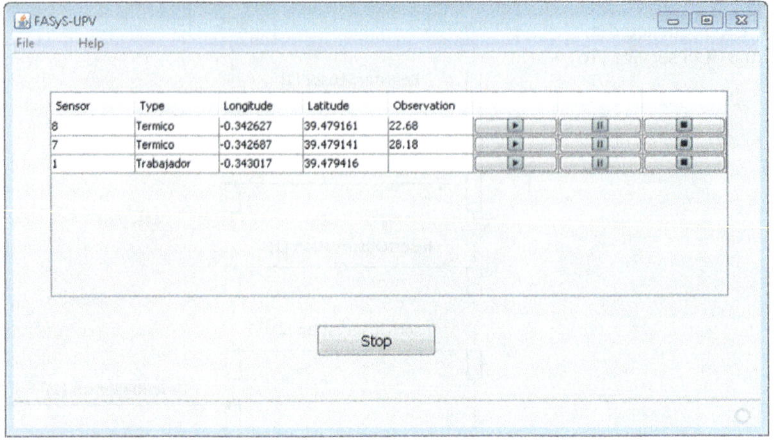

Fig. 5 Simulator HMI [16]

One can see the Sensor ID, the (physical) property that is monitored, the location (longitude and latitude) and the current measurement. Management and execution of a sensor is independent from the others so each sensor may independently be stopped, paused or played.

The process of loading data is performed by means of a configuration file. All sensors listed in the configuration file are displayed in the GUI. Additionally, the user may add new sensors and the configuration file is updated. When the simulator is started all displayed sensors start sending measurements to the SOS, and the columns in the GUI are updated when any of the longitude, latitude or observation value changes.

The data generated by the simulator changes depending on the simulation mode. As commented in the previous section, it is possible to introduce anomalies to simulate extraordinary situations, such as a fire in the case of a temperature sensor or a loud noise in the case of a sound pressure sensor. One can also define the error probability of the sensor, following a particular distribution. Or one can define a loss probability that represents lost packets that will not reach the SOS.

Data loss and delay may not represent a serious problem if measurements and observations are sent frequently, as the Event Processing module may estimate the missing values.

Sensor observations are stored in the SOS server. Later, a user application reads all data entered into the SOS by the simulator and represents it on a graphical interface. Figure 6 shows a screenshot of the main screen, with some present sensors in the scenario with particular areas where the risk level is being continuously monitored. Such level is defined as the maximum risk level of any sensor (including workers) within the specified area. The GUI also provides mechanisms to access all the information stored and related to a sensor, by just clicking on each displayed sensor (identification, location, related alarms, etc.) and even real-time data (as measurements) are inserted in the SOS (see Fig. 7).

Fig. 6 FASyS management HMI [16]

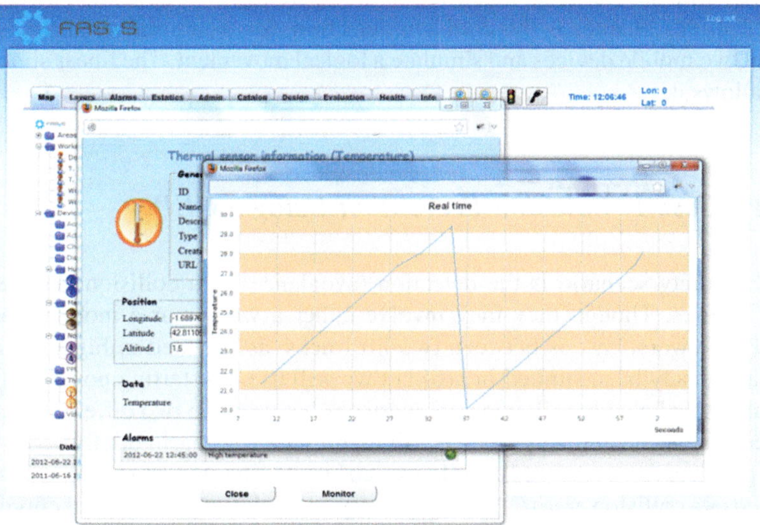

Fig. 7 FASyS sensor information detailed view [16]

5 Use Case

5.1 Factory of the Future (FoF)

Safety technologies for industrial environments have evolved considerably in recent years, but there are still risks related to the worker's safety and health. In this context, the European Factories of the Future (FoF) [30] concept focuses on the development and integration of engineering technologies, Information and Communication Technology (ICT), and advanced materials for adaptable machines and industrial processes. In this new framework, workers represent an even more important asset for the manufacturing competitiveness and productivity, and all necessary actions must be done to improve their health and safety in their working environment. All what happens in the factory, from ambient levels, to the position of all elements should be tracked and potential risks should be anticipated through automatized preventive actions.

The proposed use case belongs to a Spanish FoF project named FASyS [31]. The project was related with the development of a large wireless sensor system in order to provide safety to the workers. The project targeted a large number of use cases associated with several risks identified in the operating environment. The use case analyzed in this chapter is the collision detection, for which the simulator should generate two mobile devices and simulate a logical movement. The linear simulation model allows it.

5.2 Test Case 1: Collision Prevention Using a CEP

A typical safety scenario is the detection (avoidance) of a collision between two mobile entities. Though they may involve either a worker or a mobile machine, probably the worst case is between two lift trucks, as depicted in Fig. 8. This scenario has already been studied in [16] but we will use it a starting point for further experiments and also for allowing comparisons between the two different scenarios analysed in this chapter. Anyway we will highlight the scenario briefly.

The problem can be defined as follows (see Fig. 8). The crossing of two corridors is identified as a collision area (CA) in the factory and is determined by four coordinates (x_1, x_2, y_1, y_2). Lift trucks *a* and *b* are equipped with sensors that provide their position (pos_a, pos_b) periodically (T_a, T_b). For simplicity, we will assume uniform speed for the lift trucks, thus it is easy to obtain the corresponding speeds (s_a, s_b) from two consecutive positions. Depending on the distance to the crossing, the lift truck drivers may be alerted or not as they may be in risk. This is denoted by $x_R(sa)$ and $y_R(sb)$ respectively, and is represented by red dashed lines in Fig. 8. If both lift trucks a and b have already crossed the red lines, then there is a real risk for collision that must be avoided. Note that $x_R(sa)$ and $y_R(sb)$ depend on the speed of each lift truck, and is in line with traditional traffic regulations for calculating the braking

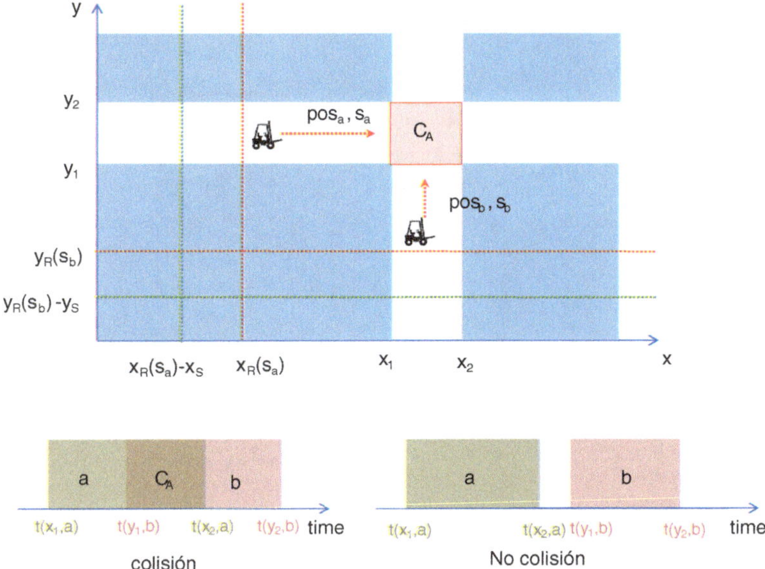

Fig. 8 Collision detection (test case) [16]

time of vehicles. Thus our system has to alert both drivers only when both lift trucks have crossed their red (risk) lines. Otherwise there will be no collision, as depicted in Fig. 8 in the timeline.

In order to monitor such situation, the Control Center (CC) must keep track of the position of each lift truck and act accordingly. The scenario is depicted in Fig. 9 for vehicle a. It inserts its position in the SOS every Ta seconds. The control center reads from the SOS every T_{read} seconds and establishes whether it has to alert lift truck a (and also vehicle b). Alerting both vehicles takes some time (T_{send}) and an acknowledgment (Tack) from each vehicle. For the factory under consideration, we have performed several tests and the mean time ($T_{send} + T_{ack}$) is 200 ms, with a variability of 50 ms. If the CC performs correctly (successfully), the driver is alerted Δx meters before crossing the risk line (depicted as a green lift truck in Fig. 9). Note that the driver may be alerted after crossing the risk line (depicted as a red lift truck in Fig. 9) and therefore exist a potential risk. However, it is possible that the other driver (b) is alerted in time and the collision risk significantly decreases.

In order to avoid alerting drivers *a* and *b* too late, there is a safety distance (x_S and y_S). The CC controls if vehicles *a* and *b* cross the safety distance (instead of the risk lines) in order to alert them shortly before they cross the risk lines. The safety lines are depicted in Fig. 8. The experiment performed in this chapter consists in checking if collision between lift trucks can be avoided or not. For this issue, we have used our simulator for different speeds (s_a and s_b) in order to check the maximum speed under which the collision risk is detected successfully. The results are presented in Table 1 for different speed values. As can be seen, if both lift trucks move with a

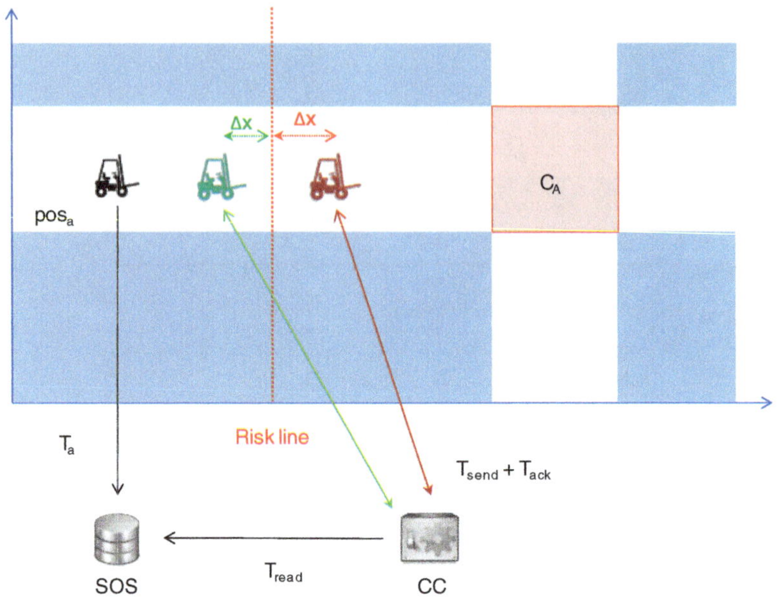

Fig. 9 Tracking of lift trucks (test case) [16]

Table 1 Simulation results for the testing scenario [16]

	Speed (s_a)	$\Delta x (m)$	Speed (s_b)	$\Delta y (m)$	Result
1	18.96	0.65	22.92	1.82	Success
2	18.96	0.73	24.45	0.62	Success
3	18.96	0.67	29.03	1.09	Success
4	21.33	0.35	33.62	0.47	Success
5	24.38	1.15	33.62	0.61	Success
6	28.44	1.48	39.74	0.71	Success
7	34.13	0.88	33.62	0.36	Success
8	34.13	0.29	50.44	−2.46	Partially Success
9	42.67	−6.28	39.73	0.23	Partially Success
10	42.67	−4.06	50.44	−0.28	Failure

speed below 42 km/h (rows 1–7), both drivers are alerted before reaching the defined threshold (risk line) and they will stop without any danger. Even if one of the drivers moves with a higher speed (rows 8, 9), the other driver is alerted before the risk line and there may be no accident (partially success).

For higher speeds on both vehicles (row 10), drivers will be alerted after the risk line and it is possible that an accident may happen as both lift trucks enter the collision area (CA).

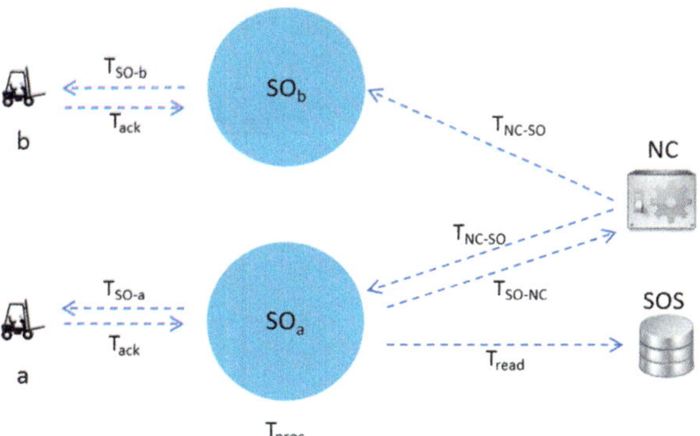

Fig. 10 Outline of events

Data loss has been set to 5 %. However, for linear movements and a low loss rate, it is not relevant if some packets (*updateSensor*) are not sent to the SOS, as the values can be estimated.

5.3 *Test Case 2: Collision Prevention Using SOs*

The employed testing scenario is the same that is designed in the first case. There are two lift trucks which move through two corridors having an area of collision within a factory.

Initialization

In general terms, any real time distributed system requires synchronization to perform actions and check status within particular timeframes. The factory, or at least some processes within it, must be considered a real time system. Initially all SO's clocks are synchronized with a time server (via NTP, Network Time Protocol) to allow an accurate calculation of collisions and establish routes. Each SOs knows beforehand the upcoming route that its corresponding vehicle will follow. This allows detecting in advance the potentially risky crossings (the lift truck requests the CRR for a factory map) and subscribing to the NC only to those events (corresponding crosses) that interest each vehicle. We assume that each vehicle insert its position in the SOS periodically; for vehicles *a* and *b* we refer to $T_{\text{InsertData}}$ (see Fig. 11).

Events

The first vehicle (say vehicle *b*) entering in the risk area of a crossing generates that its associated SO sends an alert to the notification center (NC). This action is not explicitly represented in Figs. 10 and 11 as the total amount of actions may difficult the understanding. When the other lift truck (vehicle *a*) enters in the same risk zone the associated SO detects this situation by a request to the SOS about the position of the vehicle (T_{read}) and after processing this piece of information (T_{proc}, and then it sends an alert to the NC too ($T_{\text{SO}\rightarrow\text{NC}}$). Immediately the NC communicates to all SOs subscribed to the event (in the same crossing) to avoid a collision ($T_{\text{NC}\rightarrow\text{SO}}$).

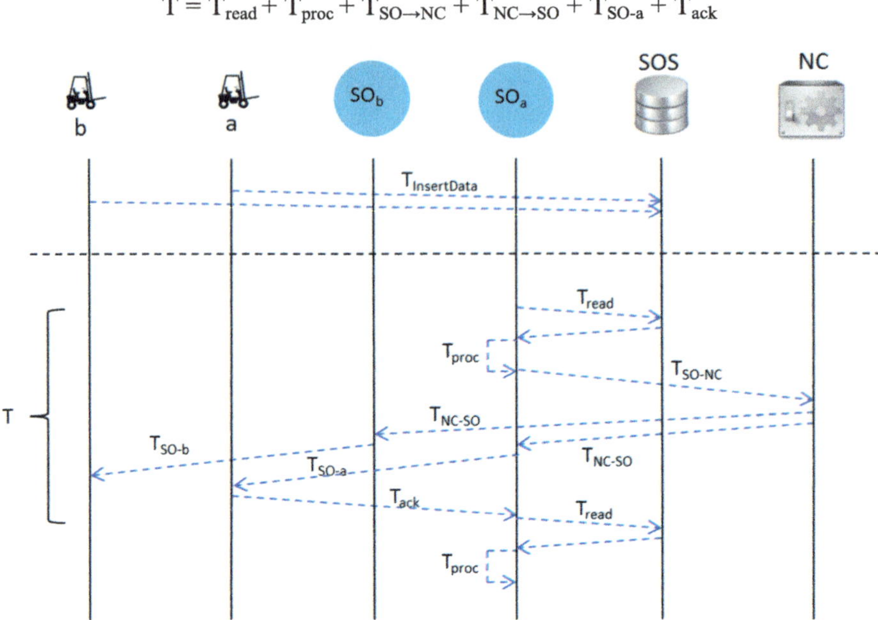

Fig. 11 Sequence diagram of the use case

Finally the SO notifies its corresponding vehicle to stop (T_{SO-a}, T_{SO-b}) and waits for a response (T_{ack}). Therefore the total amount of time for the vehicle *a* can be estimated as follows (see Fig. 11):

$$T = T_{read} + T_{proc} + T_{SO\rightarrow NC} + T_{NC\rightarrow SO} + T_{SO-a} + T_{ack}$$

After receiving the response of the vehicle, the SO requests the position to the SOS again to know where the vehicle has stopped and evaluates the result.

Results

To get these results we have simulated a factory with 25 vehicles and an area of 2500 m^2. Random routes are established for vehicles through the factory corridors. If there is a risk of collision we denote it as RC. We have established three fixed values of RC per hour: 5, 10 and 20 during a whole day (24 h). The number of vehicles (25) and the variability of RCs (between 5 and 20 per hour) seems realistic for the factory under consideration.

As one can see in the Table 2, when the number of RC increases, the maximum speed that can be used by the lift trucks to safely detect the collision decreases; this is mainly originated by the increase in the number of messages exchanged (to/from the SOS and to/from the NC). Therefore it increases the resulting time detection. For small RC values vehicles can go to a higher speed.

Table 2 Simulation results in the SO scenario

RC (ph)	Maximum speed (km/h)
5	44
10	35
20	27

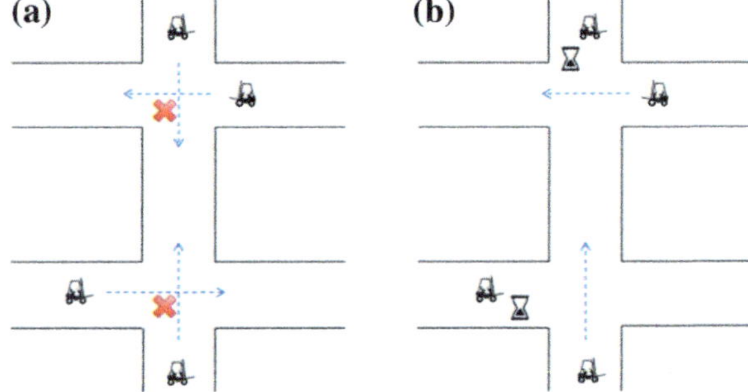

Fig. 12 Factory representation without and with scheduler

Although is difficult to compare both scenarios (test cases), because in the first one the CEP is physically in a single computer and in the second one the smart objects are distributed, we can qualitatively estimate that the final time is slightly higher. The processing time in the SO is significantly low compared to a CEP, as the SO only cares for a single vehicle whereas the CEP (test case 1) cares for the whole factory. However the CEP does not require a NC as it interacts directly with the WSNs; the SOs, on the contrary, require the NC to exchange notifications. Furthermore, the CEP can query the SOS to retrieve more information on a single message whereas each SO requires single (and simpler) messages. This increases the total amount of messages exchanged and thus the resulting time.

As the SOs are independent objects, the system is decentralized, so if one of them goes down it does not imply the failure of the whole system. The use of cloud computing (self-healing) mechanisms also helps in detecting failures and recovering immediately. However as there are multiple virtual objects there is a need for more powerful computers in the factory.

Another experiment is related to the scheduler, which reduces the effective CRs up to 150 %. In this case all SOs co-operate to establish optimal routes depending on the number of vehicles in the factory. If there is a CR, timeouts are set to everyone so that they can drive safely (see Fig. 12). Another possibility consists in changing (scheduling) the speeds in advance so that lift trucks do not collide, but this approach has not been studied here. In the end one gets a faster and more efficient system, and also requires less power consumption (less messages and less processing power).

6 Conclusions and Future Work

The system described in this chapter is a useful tool that allows analysing and assessing safety scenarios for workers and machines in industrial environments. Though the analysis has focussed on a specific scenario (collision detection and avoidance) it is easily extendable to other environments and scenarios (e.g. smart cities, e-health, etc.).

The presented system is able to avoid collisions between automatic machineries or lift trucks in a factory. To achieve this goal several components are needed: (i) WSNs to sense the environment (ii) a SOS to allow interoperability between WSN data, (iii) a sensor simulator to replicate specific behaviours from different sensors, (iv) a cloud computing environment to virtualize smart objects and provide processing capability and intelligence to WSN, (vi) a CEP to centralize some processes and compare to smart objects and (vii) an HMI to provide a graphical monitoring platform. Using the data that the WSN inserts into the SOS, the SOs can calculate in real time when a lift truck incurs in a risk of having an accident (collision) and automatically alerts it. The use of virtual smart objects provides more processing power to the WSNs, and also a private information management.

Smart objects provide a distributed management platform compared to a centralized CEP and some security concerns can be better encapsulated (e.g. personal data). The modularity of the whole system is also enhanced as new SOs can be added from different vendors with simple configuration through the SOS and NC. However cloud computing infrastructures within a factory requires a significant initial (economical) investment that has to be considered. However, energy consumption efficiency is one of the cloud computing features and guides to the concept of a sustainable factory.

Another application in the control center is the HMI where sensors in the factory are displayed and monitored. If an alarm associated to a sensor is detected, it is immediately depicted in the HMI. The HMI is layer based and allows analysing individual risks of a factory (e.g. collision detection)

As further work it would be interesting to analyse more risk scenarios within the factory, comparing results obtained from a CEP with those obtained with SOs. The correct management of SOs as they increase is also a research topic within the scope of cloud computing techniques. Finally new models and evaluation scenarios are envisioned, considering multiple sensors from different types (e.g. cameras, chemical sensors, etc.).

Acknowledgments This work has been partially funded by the Spanish Ministry of Industry under the project FASyS (Absolutely Safe and Healthy Factory) grant number CENIT 2009-1034.

References

1. Kortuem, G., Kawsar, F., Fitton, D., Sundramoorthy, V.: Smart objects as building blocks for the Internet of things, internet computing. IEEE **14**(1), 44–51 (2010)

2. Schreiber, D., Luyten, K., Mühlhäuser, M., Brdiczka, O., Hartman, M.: Introduction to the special issue on interaction with smart objects. Trans. Interact. Intell. Syst. **3**(2), 6 (2013)
3. Fortino, G., Guerrieri, A., Russo, W.: Agent-oriented smart objects development. In: Proceedings of IEEE 16th International Conference on Computer Supported Cooperative Work in Design (CSCWD), pp. 907–912 (2012)
4. Fortino, G., Guerrieri, A., Lacopo, M., Lucia, M., Russo, W.: An Agent-based Middleware for Cooperating Smart Objects. In: Highlights on Practical Applications of Agents and Multi-Agent Systems, Communications in Computer and Information Science (CCIS), Vol. 365, pp. 387–398, Springer (2013)
5. Fortino G., Guerrieri A., Russo W., Savaglio C.: Middlewares for Smart Objects and Smart Environments: Overview and Comparison, in Internet of Things based on Smart Objects: technology, middleware and applications, Springer Series on the Internet of Things (2014)
6. Hartmann, M., Schreiber, D., Mühlhäuser, M.: Workshop on interacting with smart objects. In: Proceedings of the 16th International Conference on Intelligent User Interfaces (IUI '11), pp. 481–482, New York (2011)
7. Montenegro, G., Kushalnagar, N., Hui, J., Culler, D.: RFC 4944: Transmission of IPv6 Packets over IEEE 802.15.4 Networks, Internet Engineering Task Force (2007)
8. Konstantopoulos, C., Pantziou, G., Gavalas, D., Mpitziopoulos, A., Mamalis, B.: A Rendezvous-based approach enabling energy-efficient sensory data collection with mobile sinks. IEEE Trans. Parallel Distrib. Sys. **23**(5), 809–817 (2012)
9. Liu, A.F., Ma, M., Chen, Z.G., Gui, W.: Energy-hole avoidance routing algorithm for WSN. In: Proceedings of the Fourth International Conference on Natural Computation (ICNC'08), pp. 76–80 (2008)
10. Zhang, W., Wang, Y., Ma, Y.: Research of WSN routing algorithm based on the ant algorithm. In: Proceedings of the 9th International Conference on Electronic Measurement and Instruments (ICEMI'09), pp. 422–426 (2009)
11. Ji, S., Pei, Q., Zeng, Y., Yang, C., Bu, S.: An automated black-box testing approach for WSN security protocols. In: Proceedings of the 7th International Conference on Computational Intelligence and Security, pp. 693–697 (2011)
12. Clark, J., Daigle, G.: The Importance of Simulation Techniques in ITS Research and Analysis. In: Proceedings of the 29th Conference on Winter simulation (WSC '97) (1997)
13. Wu, H., Luo, Q., Zheng, P., Ni, L.M.: VMNet: realistic emulation of wireless sensor networks. IEEE Trans. Parallel Distrib. Syst. **18**(2), 277–299 (2007)
14. The OGC Sensor Web Enablement (SWE), Open Geospatial Consortium (OGC). http://www.opengeospatial.org/ogc/markets-technologies/swe/ (2013)
15. Sensor Observation Service (SOS), Open Geospatial Consortium (OGC). http://www.opengeospatial.org/standards/sos (2013)
16. Giménez, P., Molina, B., Palau, C.E., Esteve M.: Sensor web simulation and testing for the IoT. In: IEEE International conference on Systems, Man, and Cybernetics (IEEE SMC 2013), Manchester, October 2013
17. McFarlane, D., Sarma, S., Chirn, J.L., Wong, C.Y., Ashton, K.: The intelligent product in manufacturing control and management. In: 15th Triennial World Congress IFAC, Barcelona, Spain (2002)
18. Bajic, E.: Ambient services modeling framework for intelligent products, UbiComp 2005. In: Workshop on Smart Object Systems, Tokyo Sept 2005
19. Clayman, S., Galis, A.: INOX: a managed service platform for inter-connected smart objects, ACM CoNext 2011. IoTSP workshop, Tokyo, December 2011
20. Bajic E.: A service-based methodology for RFID-smart objects interactions in supply chain, Int. J. Multimedia Ubiquitous Eng. **4**(3), 37–56 (2009)
21. Tarazona, G.M., Espada, J.P., Nunez-Valdez, E.R.: Using collaborative virtual objects based on services to communicate smart objects. In: Seventh International Conference on Innovative Mobile and Internet Services in Ubiquitous Computing (IMIS), pp. 456–461 (2013)
22. Rao, B.B.P., Saluia, P., Sharma, N., Mittal, A., Sharma, S.V.: Cloud computing for internet of things & sensing based applications. In: Sixth International Conference on Sensing Technology (ICST), Kolkata (2012)

23. Zhou, J., Leppanen, T., Harjula, E., Ylianttila, M., Ojala, T., Yu, C., Jin, H., Yang, L.T: Cloud things: a common architecture for integrating the internet of things with cloud computing. In: Proceedings of the 2013 IEEE 17th International Conference on Computer Supported Cooperative Work in Design (CSCWD), Whistler (2013)
24. North Sensor Web Community. http://52north.org/communities/sensorweb/ (2013)
25. Levis, P., Lee, N., Welsh, M., Culler, D.: TOSSIM: accurate and scalable simulation of entire TinyOS applications. In: Proceedings of the 1st International Conference on Embedded Networked Sensor Systems (2003)
26. Girod, L., Elson, J., Stathopoulos, T., Lukac, M., Estrin, D.: Emstar: a software environment for developing and deploying wireless sensor networks. In: Proceedings of the 2004 USENIX Technical Conference (2004)
27. Downard, I.: Simulating sensor networks in NS-2, NRL/FR/5522-04-10,073 (http://cs.itd.nrl.navy.mil/pubs/docs/nrlsensorsim04.pdf) (2004)
28. Varga, A., Hornig, R.: An overview of the OMNeT++ simulation environment. In: Proceedings of the 1st International Conference on Simulation Tools and Techniques for Communications, Networks and Systems (Simutools '08) (2008)
29. Zeng, X., Bagrodia, R., Gerla, M., GloMoSim: a library for parallel simulation of large-scale wireless networks, Parallel and Distributed Simulation, 1998. PADS 98. In: Proceedings of Twelfth Workshop pp. 154, 161, 26–29 (1998)
30. Research & Innovation Industrial Technologies, Public Private Partnerships in Research, Factories of the Future. http://ec.europa.eu/research/industrial_technologies/factories-of-the-future_en.html (2013)
31. Absolutely Safe and Healthy Factory (FASyS) project. http://www.fasys.es/en/ (2013)

The Cloud of Things Empowered Smart Grid Cities

Stamatis Karnouskos

Abstract The emergence of the Smart Grid era fuels a new generation of innovative applications and services that are built upon fine-grained monitoring and control capabilities pertaining the underlying infrastructure, such as that of future Smart Cities. Collection, processing and analytics on the Internet of Things of massively generated data, as well as potential management functions will emerge; therefore making a reality informed real-time decision making as well as its enforcement in a timely manner over complex infrastructures. The prevalence of the Cloud and its services, can very well complement the Internet of Things when it comes to massive data management, giving rise to the Cloud of Things (CoT). For the next generation applications, the CoT can enable access to generic multi-modal energy services, on-top of which development of more sophisticated solutions can be realized. We depict here such Smart Grid services for the Smart City of the future, as well as experiences from their realization.

1 Introduction

We witness a revolution that capitalizes on the Internet as well as the prevalence of networked (embedded) devices (ranging from simple ones such as sensors to complex systems) in order to empower a new generation of innovative anytime–anywhere services. The "Internet of Things" (IoT) [13] revolution enables unprecedented interconnection of networked embedded devices that further blur the line between the real and virtual world. The upcoming industrial revolution expectation heavily depends on these smart devices [14], while their benefits can become tangible in key innovation and economic growth areas such as Smart Cities, energy, industrial automation, health, aviation [1].

S. Karnouskos (✉)
SAP Research, Vincenz-Priessnitz-str.1, D-76131 Karlsruhe, Germany
e-mail: stamatis.karnouskos@sap.com

G. Fortino and P. Trunfio (eds.), *Internet of Things Based on Smart Objects*, Internet of Things, DOI: 10.1007/978-3-319-00491-4_7,

Predictions of 20–50 billions of connected devices by 2020 exist [28], while according to Gartner [35] "...over 50 % of Internet connections are things. In 2011, over 15 billion things on the Web, with 50+ billion intermittent connections. By 2020, over 30 billion connected things, with over 200 billion with intermittent connections. Key technologies here include embedded sensors, image recognition and NFC". These highly interconnected devices constitute the backbone of modern critical infrastructures, a representative example of which is the energy and more specifically the emerging Smart Grid [12, 17].

Integrating, interacting and meaningfully taking advantage of the benefits of infrastructures composed of vast numbers of such smart devices, is not trivial and several considerations need to be tackled. Integration requirements [21] reveal the direct [19] or via middlewares [9] coupling of smart devices such as sensors and actuators [31] with other systems and services. Even with the basic aspects of integration addressed, more complex ones at system of systems level [29], such as security, trust, privacy, reliability, management, etc., may become more critical [17], especially in the light of impact in real-world (due to the strong coupling of these devices for monitoring and control purposes in the physical world).

The Smart Grid vision [7, 8] goes beyond current efforts for smart metering (which focus mostly on fine-grained monitoring) and in the long run promises real-time decision making which may provide innovative solutions in energy management, sophisticated demand-response (DR) and demand-side management (DSM), optimal resource usage, reliability and security, new business opportunities etc. [23, 33, 39]. The core of Smart Grid depends on the usage of modern information and communication technologies [26, 43] that will enable real-time bidirectional communication with all (old and new) participating stakeholders. This will lead to the enhancement of existing processes but more importantly will enable a new generation of cross-layer collaborative services and the realization of applications that are not possible or affordable today [15].

The Smart Grid and its services are seen as an integral part of the Smart City of the future. Several ongoing efforts [10] strive towards capitalizing on the hyper-connected information infrastructure [26] and the collaboration [15] among the things, services, and systems expected to exist in the Smart City. Examples of them include real-time multi-channel energy monitoring [24], better energy coordination to take advantage of excess renewable energy or minimize costly peaks, usage of existing infrastructure such as the public lighting system for energy management and additional revenue generation [23, 32], engagement of electric vehicle fleets as well as other sources that could act as flexible energy storage etc.

Increasingly it becomes evident that the capability of real-time monitoring and management offered by the Internet of Things can have a significant impact on the Smart Grid [43] and its applicability in Smart City scenarios. However, the network effects [6] can only be capitalized upon, if an open infrastructure is in place where its layers can evolve independently. In this work we focus on shedding some light on emerging trends and the game-changing aspects they bring. As a example case, we will refer to the efforts within the NOBEL project [30] which has prototyped such an energy service-based infrastructure [18, 20, 22–25] and point out some key aspects for the future Smart Grid City.

2 The Cloud of Things Empowered Smart Grid

The prevalence of increasingly intelligent networked devices and systems in business and private homes sets the basis for a high degree of energy monitoring. This coupled with open formats for describing the data and exchanging it with other systems will enable their massive processing and empower sophisticated management mechanisms. Although at the "edge" of IoT multiple communication technologies might be used (e.g., ZigBee, Bluetooth etc.), at some level while the information flows towards Internet connected systems. IP support is offered nowadays at device level [19, 41], even when we speak about very resource constrained devices. As an indicative example for the latter we point to GreenWave Reality's (www.GreenWaveReality.com) WiFi-aware light bulbs that can be controlled by other devices such as smartphones.

Although today "smart meters" still refer to the enhancement of the legacy meters with modern Information and Communication Technologies (ICT), the prevalence of IoT means that probably any kind of networked device be considered as "smart meter" as it will be able to offer monitoring and control capabilities including information on the energy consumption and/or production. It may even go further offering additional info on current state and potentially task-specific energy signature adjustment capabilities for appropriate energy-wise task rescheduling. This is potentially a game-changer, especially in highly-automated environments, since now we will be able to know at very fine-grained level energy information that can flow into enterprise systems, calculate accurately the energy impact of business processes and optimize them dynamically considering a wide range of criteria such as environmental or financial impact etc.

Figure 1 depicts the paradigm change and the transition towards a Cloud-based IoT i.e. the Cloud of Things (CoT) [20] when considering the Smart Grid. Appliances (equipped with networked embedded systems) that are able to communicate (mostly) wireless, will be able to integrate, interact and cooperate [15] in, both P2P manner as well as via Cloud-based auxiliary services. This indicates that the "edge" devices which are usually resource-constrained, can now complement their capabilities with much more "powerful" ones running on the Cloud, thereby enhancing their own functions. Being able to attach to global infrastructure services and bidirectionally interact with them will boost the information exchanged among the virtual and physical world and will greatly benefit applications depending on their fusion.

The capability of not only offering the information acquired (provisioning) but also consuming information that may enhance their own operations leads to the enhancement of IoT edges (that were hardly possible before due to the increased resources required or a system-view not available at the edge), and optimisations at system level e.g. in a building, an enterprise, or even a smart grid city. A typical example of information provisioning would be the real-time measurements offered

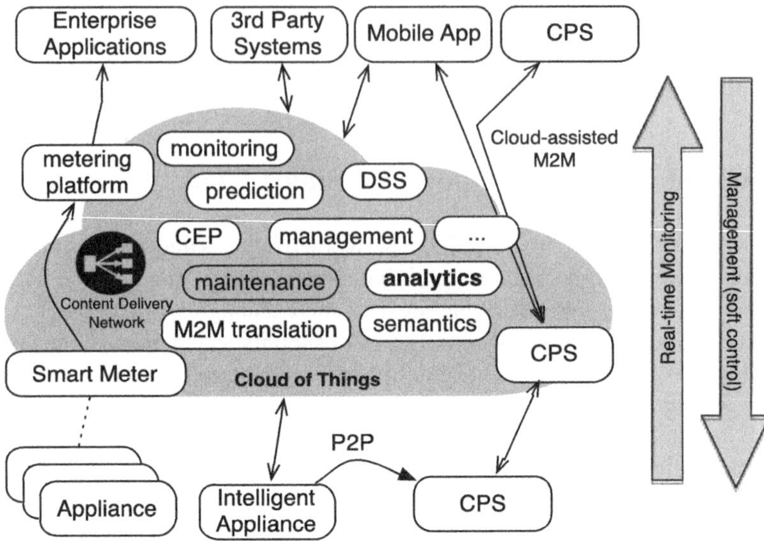

Fig. 1 The Cloud of Things in the Smart Grid context

by "edge" devices e.g. to a Cloud-based monitoring service. These can be energy measurements as well as device operation measurements which due to the rich Cloud resources available, can be analysed and decisions can be made for device specific actions (e.g. remote maintenance) as well as global actions, e.g. massive control of the consumption or generation systems. This enables new highly-scalable approaches that were previously not possible or too costly to realize; for instance avoiding blackouts by massively reducing energy consumption or rescheduling or adjusting specific categories of non-critical devices (e.g. household appliances) while giving higher priority to sensitive locations such as hospitals.

As discussed, the benefit of utilizing the Cloud of Things is that additional capabilities potentially not available at resource constraint devices can now be fully utilized taking advantage of Cloud characteristics such as virtualisation, scalability, multi-tenacy, performance, lifecycle management etc. The manufacturer for instance can use such Cloud based services in order to monitor the status of the deployed appliances, make software upgrades to the firmware of the devices, detect potential failures and notify the user, schedule proactive maintenance, get better insights on the usage of his appliance and enhance the product etc. At smart grid city level, seamless monitoring and adjustment can be done for public infrastructure to comply with the goals pursued.

The fusion of information from the business (virtual) and physical (real) world are key in achieving innovation and efficiency. The Cloud of Things makes it possible that information, generated at highly distributed points in the real world, is now available offering high visibility to aspects that were not measurable at affordable costs before. Such massive data often referred to as "Big Data" can now be collected and analysed in the Cloud, and potentially enhance decision making processes. Customized analytics can be offered to a plethora of end-devices in various forms, e.g. of the individual user or in custom-aggregated form via content-delivery-networks. The latter has the potential to empower a new generation of services that are more accurate, near real-time, and can be applied at large scale. A typical example scenario in Smart Cities would be the monitoring of Key Performance Indicators at city- or neighbourhood-wide level that pro-actively enables tackling of issues and also enhances city planning processes.

The "Big Data" acquired can be enriched with context-specific and system-wide aspects as well as business relevant information in the Cloud and enable us (e.g. through analytics) to better understand the physical world, its processes, the impact on the business side and eventually take more informed decisions. Although "Big Data" existence and analytics don't necessarily guarantee better decisions, potential new insights that may be acquired, may materialize to more effective problem tackling and business advantages. As an example, in the Smart Grid, analytics empower scenarios [10] of grid infrastructure optimization, energy management scenarios (such as demand-response schemes) with participation of residential prosumers (energy producers and consumers), energy trading, better planning of energy infrastructure in cities etc. Big data analytics is seen also as the key into understanding complex system of systems, such as the emerging Smart Cities. For instance the SmartKYE project (www.smartkye.eu) aims at enabling municipalities to better understand and manage aspects of a Smart City via a business cockpit, by relying on analytics that can be queried over a distributed infrastructure of energy management systems (EMS) and their combination with business information as well as goals.

The existence of the Cloud of Things, will constrain the need for on-premise middleware and proprietary solutions. New service providers will flourish and value added services will be created such as real-time energy monitoring, real-time billing, direct asset management, customized information services, marketplace interaction etc. This is a significant shift for the energy domain, as we move away from heavy-weight monolithic applications towards much more dynamic, up-to-date and interactive ones utilizing local capabilities. By increasing visibility via near real-time acquisition and assessment of the energy related information, providing analytics on it and allowing selective management [40], we expect the emergence of a new generation of customized energy efficiency services offered potentially even at smart grid city level [22, 25].

3 Envisioned Smart Grid City Energy Services

Although the IoT and is an area of high interest in industry, we have still to reach common understanding even for matters such as what a smart meter should offer, i.e. "basic services" to be offered by all smart meters or even data formats for open communication of the acquired data. The same holds true for any potential "basic" management capabilities that the smart meter should have in order to be able to take the envisioned role in the Smart Grid ecosystem [20]. Generally, although there are several efforts and prototypes, the Internet of Things lacks even the basic standardized API for interaction among its components i.e. the devices themselves as well as between the devices and higher systems and services. The key message is clear: To be able to capitalize on the IoT/CoT, we need to establish a common minimal base and offer basic services upon which more sophisticated approaches can be built.

As already discussed, we expect several energy services to exist in the future Smart Grid infrastructure that can be integrated into applications and traditional systems in order to enhance their functionality [18]. These services can then be integrated into applications and other mash-up services and be part of a larger Smart City ecosystem. Examples would include:

- *Timely Energy Monitoring* is expected to be a reality, especially when considering the vast investments in smart metering projects [10]. Information acquired at several layers of the Smart Grid infrastructure can be validated and securely communicated among the different systems and the Cloud-based services at (almost) real-time tempo.
- *Fine-grained Control/Management* capabilities are expected to complement existing scenarios of adaptive management of the infrastructure. This implies potential understanding of the underlying processes to a certain extend and flexibility of the energy consumption/production that can be negotiated; hence goes beyond simple ON/OFF signals and consider the whole lifecycle of affected devices and systems as well as their operational context, involved processes and goals locally and at system level.
- *Energy Brokering* [11] may be seen as a value-added service with the help of which financial management (soft-control) can be applied to the Smart Grid infrastructure. Although this is still at early stage, the implications for applications build around it could have a significant impact on the dynamic operation of the grid, as well as the offering of new innovative services and applications for all stakeholders [18].
- *Real-time Analytics* are expected to operate on the "Big Data" provided by the plethora of Smart Grid City stakeholders. Effective assessment of data will provide new insights on the existing operational aspects and unveil optimization opportunities. Additionally, informed decision making considering real-time data on an unprecedented scale will be possible, which may lead to better decisions and future planning for Smart Cities.
- *Community Management* services will provide customized information adjusted to the goals of the specific community, e.g. a neighbourhood within a Smart City, and hence actively enable a critical mass of prosumers in the Smart City. The communityware Smart Grid [16] must support the creation of dynamic communities where the (mobile) user may connect and participate with its assets (e.g. electric vehicle, white-label appliances etc.). These communities may be motivated by several aspects, e.g. environmental, economic, social etc. Support for intra-but also inter-community collaboration is wished in order to increase network effects.

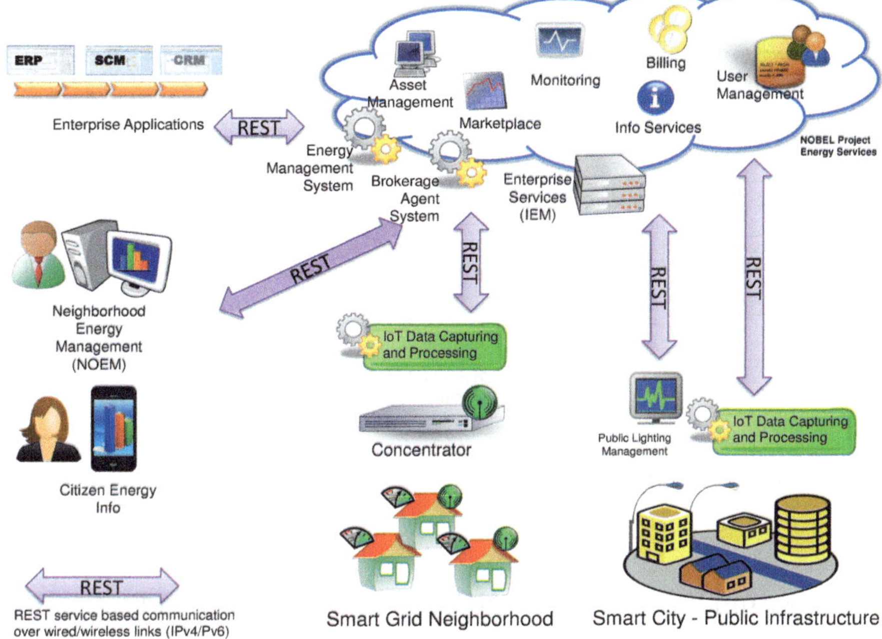

Fig. 2 Energy Service Infrastructure: the NOBEL project approach

- *Energy Application Stores* will be required to manage the large number of energy related services and applications available for the Cloud of Things enabled devices. There users/devices may automatically find, install and maintain a variety of energy related applications and services that may enhance their operational context.

As an example of emerging services the NOBEL project [30] has realized energy monitoring, management and energy brokering services (Fig. 2). Others such as BeyWatch [34], OpenMeter (www.openmeter.com), AIM [38], DEHEMS [37], BeAware [2], MIRABEL [36], ENERSip [5] and SmartHouse/SmartGrid [27] focused on intelligent device integration for energy monitoring, and complementary factors such as price-driven control, forecasting and scheduling. Similar efforts exist also in various other R&D projects [10]; however the real-time analytics, community support focus, and addressing of large-scale real-world infrastructures seem to be still at a very early stage. Recently projects such as the SmartKYE (www.smartkye.eu) were initiated with the aim to tackle the area of better decision making for municipalities based on analytics and interaction with the Smart City energy management infrastructure based on Cloud services and distributed data queries.

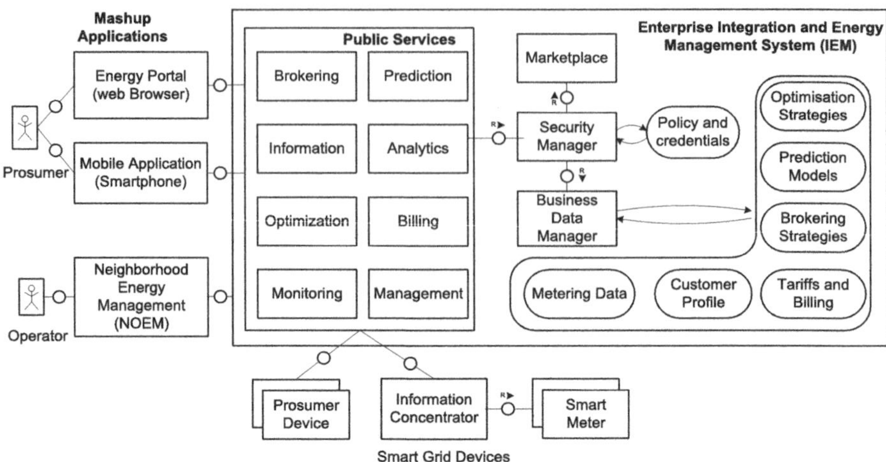

Fig. 3 Overview of the IEM architecture [22]

4 Realization Example: The NOBEL Energy Services Platform

The NOBEL [30] vision builds upon the expectation that the future energy monitoring and management system will be in close cooperation with enterprise systems. Enterprise services will integrate information coming from highly distributed smart metering points in near real-time, process it, and take appropriate decisions. This will give rise to a new generation of mash-up applications that depend on "real-world" services which constantly hold actualized data as they are generated.

The NOBEL project has taken up the challenge to design, built and pilot an open energy services platform, i.e. the Enterprise Integration and Energy Management System (IEM) [22, 25] as depicted in Fig. 3. The approach was driven by the wish to enable lightweight Internet accessible energy services for thin clients over multiple channels, thus lowering the integration and application development costs.

As proof of concept a common set of energy services has been realized, and offered to all stakeholders externally and internally to the platform via lightweight RESTful services. More specifically these were accessed by a web application, various mobile applications realized in android devices, enterprise systems and smart meters / smart meter concentrators. Based on this common set of available public APIs, a variety of services were built and piloted [24, 25].

In a Smart City, numerous systems (including potentially individual devices) may connect directly or indirectly (e.g. via gateways) to the services provided by the IEM. As seen in Fig. 3, there are several architecture parts such as the device layer, the enterprise services and end-user mash-up applications. On the IEM service layer, one can mash up services to provide customized functionalities for various applications, such as an energy portal (accessible via web browsers), mobile applications, or a neighbourhood energy management centre (NOEM) [24]. Furthermore, enterprise services process the collected data and provide advanced functionalities such as validation, analytics, and business context specific processing.

The following services have been realized and assessed [22, 25]:

- *Energy Monitoring* for acquiring and delivering data related to the energy consumption and/or production of prosumer devices.
- *Energy Prediction* for forecasting consumption and production based on historical data acquired by IEM and other third party services (e.g. weather data).
- *Management* for handling the asset, user and configuration issues in the infrastructure.
- *Energy Optimization* for interacting with existing assets of the Smart City (as a proof of concept a public lighting system [23]) and enable better usage of the energy available at neighbourhood level.
- *Brokerage* offering energy trading to all prosumers citizens who, in a stock-exchange manner, could interact via the platform and buy potentially cheaper energy or sell excess production from their photovoltaic panels.
- *Billing* which offered real-time view of the energy costs and benefits (from transactions on the Smart City energy market); hence avoiding “bill-shock” scenarios.
- *Other* value added services offering bidirectional interaction between the users and the energy provider such as notification for extra-ordinary events etc.

The RESTful web services offered by the IEM platform were designed for high performance and their API was conceived to cover the needs of the heterogeneous stakeholders accessing them. The REST adoption imposes some architectural style selections, e.g. client-server separation of concerns, stateless interactions, uniform interfaces and a layered system. An example of Smart City wide functionality shown to the city officials is depicted in Fig. 4. There a high level view of the energy production and consumption by aggregating all device measurements is provided, including some additional information such as the generation mix.

The platform (IEM) as well as example applications (e.g. NOEM) have been extensively tested and used in a pilot [24, 25] in the second half of 2012 as part of the NOBEL project pilot which took part in the city of Alginet in Spain. Data in 15 min resolution of approximately 5000 m were streamed over the period of several months to the IEM, while the IEM services were making available several functionalities ranging from traditional energy monitoring up to futuristic energy trading.

5 Considerations for Smart Cities

The hands-on experiences with the design, prototyping and piloting of the IEM and NOEM have revealed several technical aspects that should be given attention prior to productively deploying such systems [24, 25]. Apart from these however, high-level

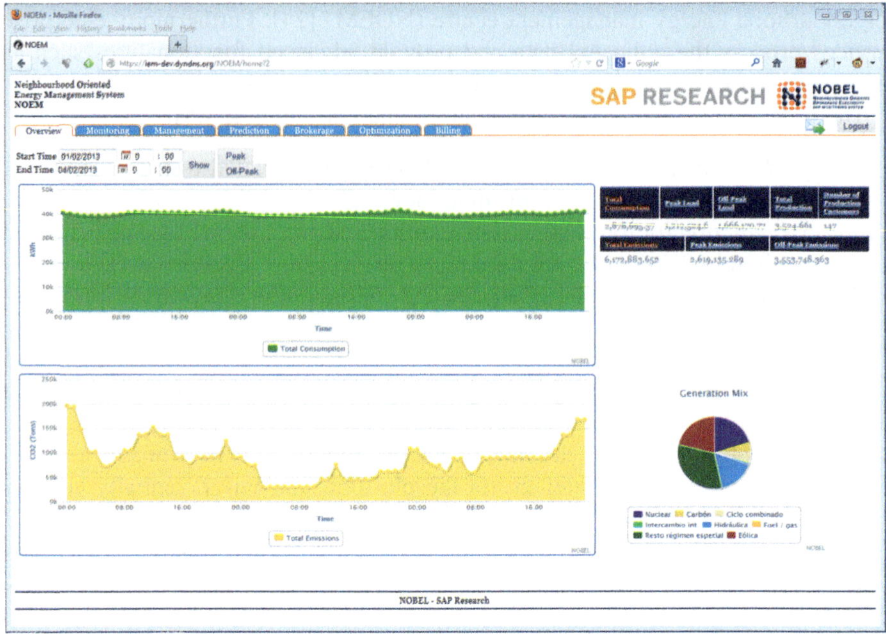

Fig. 4 Smart City cockpit showing city-wide demand, CO_2 and energy mix [24]

lessons were also acquired that could serve future Smart City stakeholders in their efforts to develop the next generation of services and applications in the Cloud of Things empowered Smart Grid.

The evolution on the infrastructure (either by retrofitting or by replacement) with the new IoT capable systems) has several challenges [4] and we will see a stepwise evolution with varying degrees of IoT-enabling technologies deployed in different parts of the Smart Cities. All these will be owned by a different stakeholders, hence the focus should not be on tightly integrated and "island", task- or process-specific only solutions, but rather on having the bigger picture in mind. The latter implies as key design criteria openness and compliance to standards, as well as focus on collaboration [31] and interaction/communication potentially in a service-oriented lightweight way. Therefore, design and deployment of Smart City wide services should be extensible and easily enable integration in future scenarios.

The Cloud of Things is seen as an integral part of future Smart Cities. Although IoT today focuses on the communication part among the different "things", the CoT makes some assumptions on the communication layer, e.g. focusing mostly on widely accepted web technologies (without excluding of course others), and integration with the Cloud in order to take advantage of its capabilities such as virtualisation, scalability, multi-tenacy, performance, lifecycle management etc. The differentiating part here is the "outsourcing" model where the edge of IoT is used for monitoring and control aspects but "heavyweight" processing can be moved from the edges to the network (i.e. the Cloud) and hence combine the best of both worlds. In the next decades with the advances in hardware capabilities of the edges, the Cloud of Things might be able to migrate closer to them (i.e. the devices) and hence it should not be treated solely with the view we have today about it i.e. running exclusively on huge centralized data-centres.

With the enhanced monitoring capabilities at IoT layer as well as the increased access and correlation with enterprise data, the era of IoT "Big Data" reaches the Smart Cities. This has profound implications on the applications and services depending not only on the processed outcome of the data but also their quality and timely acquisition. To this end, e.g. validation of data values and syntax based on model semantics, correct time-stamping, duplicate detection, security validation and risk analysis, anonymisation, data normalization, estimation of missing data, conversion to other formats or models etc. may be necessary prior to release of the data for further processing or consumption. Scalable data management [42] supporting acquisition, processing and customized consumption by a multitude of devices is key to the success.

As this is a complex system, any pitfalls may propagate and lead to high-impact system-wide problems e.g. with financial impact, prediction estimations, operational hazards etc. As the Smart City will not only monitor but also manage (control) its assets, the development of these safety-critical applications will be increasingly challenging. Since they will depend on various services (under the control of various stakeholders) it will be difficult to do systematic testing and hence safety and resilience has to be a priority [17].

The Smart City infrastructure and its services are expected to continuously evolve. Hence there is need for robust services identified as "generic" that serve the majority of applications and upon which more sophisticated interoperable [3] approaches can be built. Additionally lifecycle management has to be supported in order to enable consistent management of such a large infrastructure. In such a multi-stakeholder environment, taking also into consideration the prevalence of IoT in personal and business domains, security, trust and privacy are expected to be challenging issues. These need to be an integral part of design, implementation and deployment of any energy services and of the infrastructure they depend upon.

6 Conclusions

Once cooperative infrastructures are in place within a Smart City, a plethora of applications are expected to flourish and offer new innovative services. The latter will enable direct interaction between all Smart City stakeholders for the common benefit. Smart Cities can harness the power of the emerging IoT/CoT enabled Smart Grid and enhance their operation to better suit the needs of their citizens. Informed decisions can now be taken that rely more on actual data and up-to-date analytics, empowering decision-makers with new opportunities for better management and eradication of inefficiencies.

To harness the benefits, there are challenges at technology,business, and social levels that need to be addressed. The future Smart Cities will need to be built on principles of cooperation, openness/interoperability and trust. Extracting and understanding the business relevant information under temporal constraints and being able to effectively integrate them into solutions that utilize the monitor-analyse-decide-manage approach for a multitude of domains is challenging. The high heterogeneity of systems, models, quality of data and associated information, uncertainties as well as complex system-wide interactions, will need to be investigated to identify business opportunities and realize (business) benefits for all stakeholders. Considering also that for instance in Smart Cities much of the "Big Data" will be directly generated by or affect its citizens, data lifecycle management approaches with security and privacy aspects integrated will need to be well-thought.

Acknowledgments The author would like to thank Dejan Ilic and Per Goncalves Da Silva of SAP Research, as well as the partners of the EU research projects NOBEL (www.ict-nobel.eu) and SmartKYE (www.smartKYE.eu).

References

1. acatech. Cyber-Physical Systems: Driving force for innovation in mobility, health, energy and production. Tech. rep., acatech—National Academy of Science and Engineering. http://www.acatech.de/fileadmin/user_upload/Baumstruktur_nach_Website/Acatech/root/de/Publikationen/Stellungnahmen/acatech_POSITION_CPS_Englisch_WEB.pdf (2011)
2. Björkskog, C.A., Jacucci, G., Gamberini, L., Nieminen, T., Mikkola, T., Torstensson, C., Bertoncini, M.: Energylife: pervasive energy awareness for households. In: Proceedings of the 12th ACM International Conference Adjunct Papers on Ubiquitous Computing—Adjunct, Ubicomp '10 Adjunct, ACM, New York, pp. 361–362 (2010). doi:10.1145/1864431.1864436
3. Bryson, J., Gallagher, P.D.: NIST framework and roadmap for smart grid interoperability standards, release 2.0. Tech. Rep. NIST Special Publication 1108R2, National Institute of Standards and Technology (NIST). http://www.nist.gov/smartgrid/upload/NIST_Framework_Release_2-0_corr.pdf (2012)
4. Carosio, S., Hannus, M., Mastrodonato, C., Delponte, E., Cavallaro, A., Cricchio, F., Karnouskos, S., Pereira-Carlos, J., Rodriguez, C.B., Nilsson, O., Seppä, I.P., Sasin, T., Zach, J., van Beurden, H., Anderson, T.: ICT Roadmap for Energy-Efficient Buildings—Research and Actions. EU Project ICT4E2B Consortium. http://ict4e2b.atosresearch.eu/sites/ict4e2b.atosresearch.eu/files/page-files/13/ICT4E2B/20Forum/20Book.pdf (2013)
5. Carreiro, A.M., Lopez, G.L., Moura, P.S., Moreno, J.I., de Almeida, A.T., Malaquias, J.L.: In-house monitoring and control network for the Smart Grid of the future. In: Innovative Smart Grid Technologies Conference Europe IEEE PES, pp. 1–7 (2011). doi:10.1109/ISGTEurope.2011.6162736
6. Easley, D., Kleinberg, J.: Networks, Crowds, and Markets: Reasoning About a Highly Connected World. Cambridge University Press, Cambridge. http://www.cs.cornell.edu/home/kleinber/networks-book/ (2010)
7. European Commission. SmartGrids SRA 2035—Strategic Research Agenda: Update of the SmartGrids SRA 2007 for the needs by the year 2035. Tech. rep., European Technology Platform SmartGrids, European Commission. http://www.smartgrids.eu/documents/sra2035.pdf (2012)

8. Federation of German Industries (BDI). Internet of Energy: ICT for energy markets of the future. BDI publication No. 439. http://www.bdi.eu/BDI_english/download_content/ForschungTechnikUndInnovation/BDI_initiative_IoE_us-IdE-Broschure.pdf (2010)
9. Fortino, G., Guerrieri, A., Russo, W., Savaglio, C.: Middlewares for smart objects and smart environments: overview and comparison. Internet of Things Based on Smart Objects: Technology, Middleware and Applications. Springer Series on the Internet of Things (2014)
10. Giordano, V., Meletiou, A., Covrig, C.F., Mengolini, A., Ardelean, M., Fulli, G., Jiménez, M.S., Filiou, C.: Smart Grid Projects in Europe: Lessons Learned and Current Developments 2012 Update. Joint Research Center of the European Commission, JRC79219 (2013). doi:10.2790/82707
11. Ilić, D., Goncalves Da Silva, P., Karnouskos, S., Griesemer, M.: An energy market for trading electricity in smart grid neighbourhoods. In: 6th IEEE International Conference on Digital Ecosystem Technologies—Complex Environment Engineering (IEEE DEST-CEE), Campione d'Italia, Italy (2012)
12. Ilic, M., Xie, L., Khan, U., Moura, J.: Modeling future cyber-physical energy systems. In: Power and Energy Society General Meeting—Conversion and Delivery of Electrical Energy in the 21st Century, pp 1–9 (2008). doi:10.1109/PES.2008.4596708
13. International Telecommunication Union. ITU Internet Report 2005: the internet of things. Tech. rep., (ITU) (2005)
14. Kagermann, H., Wahlster, W., Helbig, J.: Recommendations for implementing the strategic initiative INDUSTRIE 4.0. Tech. rep., acatech—National Academy of Science and Engineering. http://www.acatech.de/fileadmin/user_upload/Baumstruktur_nach_Website/Acatech/root/de/Material_fuer_Sonderseiten/Industrie_4.0/Final_report_Industrie_4.0_accessible.pdf (2013)
15. Karnouskos, S.: The cooperative internet of things enabled smart grid. In: Proceedings of the 14th IEEE International Symposium on Consumer Electronics (ISCE2010), June, pp. 07–10. Braunschweig, Germany (2010)
16. Karnouskos, S.: Communityware smartgrid. In: 21st International Conference and Exhibition on Electricity Distribution (CIRED 2011), Frankfurt (2011a)
17. Karnouskos, S.: Cyber-physical systems in the smartgrid. In: IEEE 9th International Conference on Industrial Informatics (INDIN), Lisbon (2011b)
18. Karnouskos, S.: Demand side management via prosumer interactions in a smart city energy marketplace. In: IEEE International Conference on Innovative Smart Grid Technologies (ISGT 2011), Manchester (2011c)
19. Karnouskos, S.: Asset monitoring in the service-oriented internet of things empowered smartgrid. Service Oriented Computing and Applications, pp. 1–8 (2012). doi:10.1007/s11761-012-0102-6
20. Karnouskos, S.: Smart houses in the smart grid and the search for value-added services in the cloud of things era. In: IEEE International Conference on Industrial Technology (ICIT 2013), Cape Town (2013)
21. Karnouskos, S., Vilaseñor, V., Handte, M., Marrón, P.J.: Ubiquitous integration of cooperating objects. Int. J. Next-Generat. Comput. **2**(3), (2011)
22. Karnouskos, S., Goncalves Da Silva, P., Ilić, D.: Energy services for the smart grid city. In: 6th IEEE International Conference on Digital Ecosystem Technologies—Complex Environment Engineering (IEEE DEST-CEE), Campione d'Italia, Italy (2012a)
23. Karnouskos, S., Ilić, D., Goncalves Da Silva, P.: Using flexible energy infrastructures for demand response in a smart grid city. In: The third IEEE PES Innovative Smart Grid Technologies (ISGT) Europe, Berlin (2012b)
24. Karnouskos, S., Goncalves Da Silva, P., Ilić, D.: Developing a web application for monitoring and management of smart grid neighborhoods. In: IEEE 11th International Conference on Industrial Informatics (INDIN), Bochum (2013a)
25. Karnouskos, S., lić, D., Goncalves Da Silva, P.: Assessment of an enterprise energy service platform in a smart grid city pilot. In: IEEE 11th International Conference on Industrial Informatics (INDIN), Bochum (2013b)

26. Katz, R., Culler, D., Sanders, S., Alspaugh, S., Chen, Y., Dawson-Haggerty, S., Dutta, P., He, M., Jiang, X., Keys, L., Krioukov, A., Lutz, K., Ortiz, J., Mohan, P., Reutzel, E., Taneja, J., Hsu, J., Shankar, S.: An information-centric energy infrastructure: the berkeley view. Sustainable Computing: Informatics and Systems (2011)
27. Kok, K., Karnouskos, S., Ringelstein, J., Dimeas, A., Weidlich, A., Warmer, C., Drenkard, S., Hatziargyriou, N., Lioliou, V.: Field-testing smart houses for a smart grid. In: 21st International Conference and Exhibition on Electricity Distribution (CIRED 2011), Frankfurt (2011)
28. Lomas, N.: Online Gizmos Could Top 50 Billion in 2020. http://www.businessweek.com/globalbiz/content/jun2009/gb20090629_492027.htm (2009)
29. Maier, M.W.: Architecting principles for systems-of-systems. Systems Engineering. John Wiley & Sons, Incvol. 1, issue no. 4, pp. 267–284 (1998)
30. Marqués, A., Serrano, M., Karnouskos, S., Marrón, P.J., Sauter, R., Bekiaris, E., Kesidou, E., Höglund, J.: NOBEL—a neighborhood oriented brokerage electricity and monitoring system. In: 1st International ICST Conference on E-Energy, 14–15 Oct 2010 Athens. Springer (2010)
31. Marrón, P.J., Karnouskos, S., Minder, D., Ollero, A. (eds): The Emerging Domain of Cooperating Objects. Springer (2011). doi:10.1007/978-3-642-16946-5 NULL
32. Müllner, R., Riener, A.: An energy efficient pedestrian aware smart street lighting system. Int. J. Perv. Comput. Commun. **7**(2), 147–161 (2011). doi:10.1108/17427371111146437, http://www.emeraldinsight.com/journals.htm?articleid=1937697
33. Paetz, A.G., Becker, B., Fichtner, W., Schmeck, H.: Shifting electricity demand with smart home technologies—an experimental study on user acceptance. In: 30th USAEE / IAEE North American Conference Online Proceedings (2011)
34. Perdikeas, M., Zahariadis, T., Plaza, P.: The beywatch conceptual model for demand-side management. In: Hatziargyriou, N., Dimeas, A., Tomtsi, T., Weidlich, A. (eds.) Energy-Efficient Computing and Networking, Lecture Notes of the Institute for Computer Sciences, Social Informatics and Telecommunications Engineering, vol. 54, pp. 177–186. Springer, Berlin. doi:10.1007/978-3-642-19322-419 (2011)
35. Savitz, E.: Gartner: top 10 strategic technology trends for 2013. http://www.forbes.com/sites/ericsavitz/2012/10/22/gartner-10-critical-tech-trends-for-the-next-five-years/ (2012)
36. Šikšnys, L., Khalefa, M.E., Pedersen, T.B.: Aggregating and disaggregating flexibility objects. In: Proceedings of the 24th international conference on Scientific and Statistical Database Management, Springer, Berlin SSDBM'12, pp 379–396 (2012). doi:10.1007/978-3-642-31235-925
37. Sundramoorthy, V., Liu, Q., Cooper, G., Linge, N., Cooper, J.: Dehems: a user-driven domestic energy monitoring system. In: Internet of Things (IOT), IoT for a green Planet, Tokyo. IEEE (2010). doi:10.1109/IOT.2010.5678451
38. Tompros, S., Mouratidis, N., Draaijer, M., Foglar, A., Hrasnica, H.: Enabling applicability of energy saving applications on the appliances of the home environment. IEEE Netw. **23**(6), 8–16. (2009). doi:10.1109/MNET.2009.5350347
39. US Department of Energy. Benefits of demand response in electricity markets and recommendations for achieving them. Tech. rep., US Department of Energy. http://energy.gov/sites/prod/files/oeprod/DocumentsandMedia/DOE_Benefits_of_Demand_Response_in_Electricity_Markets_and_Recommendations_for_Achieving_Them_Report_to_Congress.pdf (2006)
40. Valocchi, M., Juliano, J.: Knowledge is power—driving smarter energy usage through consumer education. Tech. rep., IBM Institute for Business Value. http://public.dhe.ibm.com/common/ssi/ecm/en/gbe03479usen/GBE03479USEN.PDF (2012)
41. Vasseur, J.P., Dunkels, A.: Interconnecting Smart Objects with IP: The Next Internet. Morgan Kaufmann Publishers Inc., San Francisco (2010)
42. Yin, J., Kulkarni, A., Purohit, S., Gorton, I., Akyol, B.: Scalable real time data management for smart grid. In: Proceedings of the Middleware 2011 Industry Track Workshop, ACM, New York, Middleware '11, pp. 1:1–1:6 (2011)
43. Yu, X., Cecati, C., Dillon, T., Simões, M.: The new frontier of smart grids. IEEE Ind. Electron. Magaz. **5**(3):49–63 (2011). doi:10.1109/MIE.2011.942176

Trajectory Data Analysis Over a Cloud-Based Framework for Smart City Analytics

Eugenio Cesario, Carmela Comito and Domenico Talia

Abstract The chapter presents a Cloud-based framework that can be tailored to be used in different scenarios of urban planning and management occuring in Smart Cities. The focus is on the management of large-scale socio-geographic data obtained through the trajectories traced by smart objects. Our goal is to mine human activities and routines from this socio-geographic data in order to catch user's behaviour. To this aim, we introduce a methodology for trajectory pattern mining consisting in (a) finding frequent regions, more densely passed through ones, and (b) extracting trajectory patterns from those regions. Experimental evaluation shows that due to complexity and large data involved in the application scenario, the trajectory pattern mining process can take advantage from a parallel execution environment offered by a Cloud architecture.

1 Introduction

The concept of *Smart City* has been introduced as the application of ubiquitous and pervasive computing paradigms to urban spaces focusing on developing city network infrastructures, optimizing traffic and transportation flows, lowering energy consumption, and innovative services for citizens [11]. This is implemented by the integration of different control systems in commercial and public interest buildings, aimed at monitoring lighting and electricity, fire detection, video surveillance,

E. Cesario (✉) · C. Comito
ICAR-CNR, Rende (CS), Italy
e-mail: cesario@icar.cnr.it

C. Comito
e-mail: ccomito@dimes.unical.it

D. Talia
ICAR-CNR and DIMES-UNICAL, Rende (CS), Italy
e-mail: talia@dimes.unical.it

G. Fortino and P. Trunfio (eds.), *Internet of Things Based on Smart Objects*, Internet of Things, DOI: 10.1007/978-3-319-00491-4_8,

access control and public address services. Smart Cities can dramatically improve the citizen's quality of life, encourage business to invest and create a sustainable urban environment.

A fundamental aspect of a smart city is the availability of a network of intelligent devices, i.e., *smart objects*, to support the development of a number of new services. A smart object network provides a flexible approach to implement "the Internet of things," in which a large number of smart objects are connected to the Internet. By 2015, the number of smart objects connected to the Internet is expected to exceed the number of computers connected [8]. Therefore, smart object networks will play a critical role in making Smart Cities a reality.

Emerging technologies, such as wireless devices, Cloud computing and vehicular networking, promote the developments of urban computing within smart city by enabling, among other things, an alternative way for tracking and sensing exploiting people's mobile devices to track mobile events [2, 4, 5]. This leads to the generation of a large number of trajectories drawn by the users during their daily activities. Such amount of information can be analyzed to discover knowledge, i.e., patterns, rules and regularities, on the user trajectories. The basic assumption is that people often tend to follow common routes: e.g., they go to work every day by similar paths. Thus, if we have enough data to model typical behaviors, such knowledge can be used to predict and manage future movements of people [12, 16, 17].

The aim of the proposed work is twofold: (i) provide an integrated computing framework for efficiently manage fragmented socio-environmental data, with a particular focus on the urban context of cities; (ii) provide a methodology to analyze trajectories of mobile users and extract movement patterns in order to mine social and environmental behaviors.

The design and implementation of a framework analysing cross-thematic socio-environmental data coming from various application domains obtained by a variety of tools, components and services is very challenging. Managing heterogeneous data volumes while allowing inter-operability among different tools, it also needs compliance to standards. In this regard, the Cloud computing paradigm is a suitable infrastructure to fulfil most of the above requirements due to its characteristics such as high-performance computing, on-demand processing, facilitating data accessibility and storage across platforms.

Accordingly, we describe a Cloud-based architecture specifically designed for urban computing supporting smart cities. The architecture includes computing and network infrastructures integrated in a Cloud platform that interact with data source generators like sensors, smart phones and other wireless devices. The framework includes a set of services allowing to gather and collect environmental data, and to process and analyze them in order to mine social and environmental behaviors. Exploiting this information, a set of functionalities can be implemented atop the basic services allowing to improve urban planning and management [22].

In this chapter one scenario is introduced as a case study. This scenario focuses on the study of the trajectories followed by mobile devices with the aim to understand trajectories in order to catch user's behaviour. Towards this direction, the chapter introduces a methodology that first detects dense regions within a given geographical area, more densely passed through regions, and then extracts trajectory patterns from those regions. We apply the trajectory pattern extraction methodology to a real-world dataset concerning mobility of citizens within an urban area. As result of the analysis, useful insights, both in terms of common movements followed in the city and in terms of evolution of the traffic density, are provided. Experimental evaluation shows that due to complexity and large data involved in the application scenario, the trajectory pattern mining process can take advantage from a parallel execution environment as offered by the Cloud.

The chapter is structured as follows. Section 2 gives a background on both Cloud computing and Smart City. The Cloud-based architecture is introduced in Sect. 3. Section 4 presents the trajectory analysis scenario describing the trajectory pattern detection methodology together with the designed workflow. Experimental evaluation is reported in Sect. 5. Section 6 concludes the chapter.

2 Background

This section provides some background information on both Cloud computing (2.1) and Smart City (2.2) also reviewing the most relevant research and projects about the use of Cloud technology for the design and development of smart cities (2.3).

2.1 Cloud Computing

Different definitions of Cloud Computing paradigm have been conceived in literature. Some of them focus on on-demand dynamic provisioning of processing and storage resources, others emphasize the service-oriented interface and the exploitation of virtualization techniques. The National Institute of Standards and Technology (NIST) has given a complete reference definition [7]. NIST defined Clouds as follows: "Cloud computing is a pay-per-use model for enabling available, convenient, on-demand network access to a shared pool of configurable computing resources (e.g., networks, servers, storage, applications, services) that can be rapidly provisioned and released with minimal management effort or service provider interaction". Moreover, according to NIST: "Cloud model promotes availability and is comprised of five key characteristics, three delivery models, and four deployment models." The key characteristics of Clouds are: on-demand self-service, ubiquitous network access, location independent resource pooling, rapid elasticity, and pay per use.

The delivery models of Clouds are very important because they define three different types of Cloud computing systems:

Infrastructure as a Service (IaaS). The capability provided to the user is to rent computing, storage, networks, and other computing resources where the user is able to deploy and run software, which can include operating systems and/or applications. The user does not manage or control the hardware Cloud infrastructure but has control over operating environments, storage, deployed applications, and possibly selects networking components. Examples for commercial Cloud infrastructures are Amazon EC2 and Rackspace.

Platform as a Service (PaaS). The functionality provided to the user is to deploy onto the Cloud infrastructure consumer-created applications using programming languages, compilers and toolkits supported by the provider (e.g., Java, .Net). The consumer does not manage or control the underlying cloud infrastructure, network, servers, operating systems, or storage, but the consumer can control the deployed applications and possibly the application hosting environment configurations.

Software as a Service (SaaS). The capability provided to the consumer is to use the provider's applications running on a Cloud infrastructure and accessible from various client devices through a thin client interface such as a Web browser (e.g., web-based email). The consumer does not manage or control the underlying cloud infrastructure, network, servers, operating systems, storage, or even individual application capabilities, with the possible exception of limited user-specific application configuration settings

IT organizations can choose to deploy applications on public, private, community, or hybrid clouds, each of which has its trade-offs. Different Cloud deployment models are described in the following.

Public Cloud. The Cloud infrastructure is owned by an organization selling Cloud services to the general public or to enterprises. Thus, it is public because it can be rent by anyone for developing and/or running any kind of applications.

Private Cloud. The Cloud infrastructure is owned or leased by a single organization and is operated only for that organization. No public access to it is permitted. This model can be used in case of strict data privacy and/or security requirements.

Community Cloud. The Cloud infrastructure is shared by a limited number of organizations and supports a specific community that has shared concerns (e.g., goals, security requirements, policy, and compliance issues).

Hybrid Cloud. This fourth class of Cloud infrastructure is a composition of two or more Clouds (private, community, or public) that although they are unique entities, are combined together by standardized or proprietary technology that enables data and application portability (e.g., Cloud federation).

Cloud computing is the most recent results of the advancement of several computer technologies both from the hardware side, such as virtualization and multicore architectures, and from the software side like cluster computing, Grid computing, Web services, service-oriented architectures, autonomic computing, and large-scale data storage. In particular, virtualization in Cloud computing is the key element that separates system functionality and implementation from physical resources. By exploiting virtualization techniques, a Cloud infrastructure can be partitioned in several parallel virtual machines, dynamically configured according to the user requirements and devoted to run independent applications concurrently.

Virtualization separates applications from hardware and users from other users giving them the feeling that a large-scale computing infrastructure is devoted to their applications by meeting a given quality of service (QoS). Virtualization is also used to isolate applications avoiding that if one fails others can fail too. Finally, virtualization is a way to improve security and privacy of concurrent applications running on the same Cloud. Several companies set up large Cloud facilities and built programming environments where developers can program applications as Cloud software services. Just to mention some example, Amazon on his EC2 and S3 Cloud platforms implemented Elastic BeanStalk, Microsoft implemented .Net technology on Azure, Google provides the AppEngine, and VMware has Cloud Foundry. On the other side, the research community developed open source software that can be deployed and configured on servers, computer farms or data centers for implementing private, public, community or hybrid Cloud infrastructures or for inter-Cloud computing facilities. Examples of these systems are OpenNebula [19], Eucalyptus [20], OpenQRM [9], Puppet [3], and OpenStack [15]. These open source software projects are also working to systems and services that allow Cloud-to-Cloud interoperability and federation.

2.2 What is a Smart City?

The Smart City concept has been introduced as a strategic paradigm to encompass modern urban production factors in a common framework and to highlight the growing importance of Information and Communication Technologies, social and environmental capital in profiling the competitiveness of cities. The term was coined in Australia for the cities of Brisbane and Blacksbourg where the ICT supported the social participation and the accessibility to public information and services. The smart city was later evolved to an urban space for business opportunities, and to ubiquitous technologies installed across the city, which are integrated into everyday objects and activities.

From literature the term Smart City is used to describe various aspects of the city which range from Smart City as an IT-district to a Smart City regarding the education of its inhabitants. In particular, Smart City can be identified along six main axes: smart economy, smart mobility, smart environment, smart people, smart living and, finally, smart governance. In the following the various aspects will be summarized.

Smart economy includes factors all around economic competitiveness as innovation, enterprise, trademarks, productivity and flexibility of the labour market as well as the integration in the (inter-)national market.

Smart people is not only described by the level of qualification or education of the citizens but also by the quality of social interactions regarding integration and public life.

Smart governance comprises aspects of political participation, services for citizens as well as the functioning of the administration. Good governance as an aspect of a smart administration often also referred to the usage of new channels of communication for the citizens, e.g., e-governance or e-democracy.

Smart City concerns also the use of modern technology in everyday urban life. This includes not only ICT but also modern transport and communication technologies. Logistics as well as new transport systems which improve the urban traffic and the inhabitants mobility.

Smart environment is described by attractive natural conditions (climate, green space etc.), pollution, resource management and also by efforts towards environmental protection.

Finally, smart living comprises various aspects of quality of life as culture, health, safety, housing, tourism etc.

A generic multi-tier architecture for smart cities should contain the following layers:

- User layer. This layer appears both at the top and at the bottom of the generic architecture because it concerns both the local stakeholders who supervise the smart city, designing city services, and the end-users who use the smart city services also having the possibility of participate in decision making.
- Service layer, which includes all the services offered by the smart city.
- Infrastructure layer that contains network, information systems and other facilities, which contribute to service deployment.
- Data layer that maintains and made available all the information, produced and collected in the smart city.

On top of the above architecture, various services can be offered in a modern smart city, including the following ones:

- Government services concern public complaints, administrative procedures at local and at national level, job searches and public procurement.
- Business services mainly support business installation.
- Health and tele-care services offer remote support to particular groups of citizens such as the elderly, civilians with diseases etc.
- Learning services offer distant learning opportunities and training material to the habitants.
- Security services support public safety via amber-alert notifications, school monitoring, natural hazard management etc.
- Environmental services contain public information about recycling, while they support households and enterprises in waste/energy/water management. Moreover, they deliver data to the State for monitoring and for decision making on environmental conditions such as for microclimate, pollution, noise, traffic etc. (in Ubiquitous and Eco-city approaches).
- Intelligent Transportation supports the improvement of the quality of life in the city, while it offers tools for traffic monitoring, measurement and optimization.
- Smart Tourism services. The use and application of ICT technology to deliver innovative approaches and technology solutions to enhance tourism sectors technology base. As example, information about cities and powerful itinerary-planning tools

can be synced across smartphones, Google calendars, and web browsers. These tools should enable tourists to quickly find activities by type and proximity and fit those activities into personalized itineraries, all without the need for a translator, tour guide, or travel agent.

2.3 Cloud-Based Urban Computing in Smart Cities

Urban computing is emerging as a concept where every sensor, device, person, vehicle, building, and street in the urban areas can be used as a component to enable a citywide computing for serving people and cities.

More formally, urban computing can be defined as the process of acquisition, integration, and analysis of big and heterogeneous data generated by a diversity of sources in urban spaces, such as sensors, devices, vehicles, buildings, and human, to tackle the major issues that cities face, e.g., air pollution, increased energy consumption and traffic congestion.

Several Cloud enabled tools for urban planning and management in the Smart City context have been recently proposed (e.g., land, etc.). Environmental Software and Services (ESS) [5] exploits the Cloud paradigm to offer a range of services for environmental planning and management, policy and decision making, world wide. Analogously, the Environmental Virtual Observatory pilot (EVOp) [4] uses Clouds to achieve similar objectives in the soil and water domains.

The European Platform for Intelligent Cities (EPIC) [1] combines the Cloud computing infrastructure with the knowledge and expertise of the Living Lab approach to deliver sustainable, user-driven web services for citizens and businesses.

The Life 2.0 project [6] offers a set of services ranging from basic geographical positioning systems to socially networked services and to local market-based services. The project aims to provide solutions that increase opportunities for social contacts between elderly people in their local area, by providing new services for elderly people, based on the use of tracking systems and social network applications.

IBM introduced Smarter City Solutions on the IBM SmartCloud Enterprise, a public Cloud platform that includes hardware, network and storage [2]. The platform provides pay-as-you-go services for urban management within cities. Those services include application software, infrastructure, networking, systems software, middleware and maintenance.

Differently from most of the systems described above, our framework has been designed to provide general-purpose services for urban planning and management within the city context. Thus, it is not been designed for a specific application domain but it can be tailored to the different aspects concerning urban management (e.g., healthcare, transportation, tourism, etc.). Moreover, the framework has been designed as a set of modular components allowing this way easy extensibility and integration of different heterogeneous components (e.g., software, data sources, etc).

3 A Cloud-Based Architecture for Smart Cities

In this section we introduce the architecture of a Cloud framework for Smart Cities. This framework implements a set of services aming at improving the planning, managing and monitoring of activities within a urban context of cities. Such activities involve a number of aspects such as healthcare, smart transportation, smart home, smart tourism and smart public services, emergency response and public information services.

On defining an architecture for urban computing in smart cities, one has to take into account the issues and challenges arising therefrom, including the following ones:

- Urban computing involves the design and development of advanced algorithms and computational platforms to solve problems in the smart city context.
- The gathering and management of massive large-scale environmental spatial-temporal data.
- The integration of heterogeneous data coming from very different and disparate sources.
- High computing power, is required due to the large size of data sets and the complexity of basic computations.
- A secure software infrastructure is needed to access private data.
- Intelligent data analysis and decision support system: generation of environmental indicators to support policy interventions and decision-making within the urban environment.

Accordingly, the proposed architecture has been designed as a middleware substrate allowing for the integration and handling of large-scale, fragmented, cross-thematic environmental and socio-geographic data with the focus of mining human behavior from such data for urban planning and management. As such, we can envision a set of services ranging from acquiring data from disparate sources (e.g., sensors, smart phones, gps, etc) to the integration analysis and processing of such data in order to define a set of services for urban planning and management. These services include the following ones:

1. *Data Acquisition*. In this phase data are gathered from different sources including traditional database systems, flat files, web services, real-time systems, sensors networks, web portals.
2. *Data Storage*. Data acquired in the previous phase are then organized into repositories.
3. *Data Processing*. This step includes traditional data processing techniques (e.g., traditional DBMS operations) as well as new techniques and methodologies introduced to elaborate the new data models established with the diffusion of sensors and ad-hoc networks and, in general, with the emergence of the ubiquitous paradigm. For example, the use of techniques like TinyOS and TinyDB to process sensor data, new query languages and execution engines for spatio-temporal data

(as it is the case for trajectory data), algorithms to process data from maps, methodologies and algorithms for image and signal processing.

4. *Decision-making*. Optimized planning and enhanced decision-making for investment projects are critical issues for dealing with the complexities of urban contexts. The use of data mining and other artificial intelligence techniques enables users to take intelligent decisions for an action to be taken. The decision-support system (DSS) provides solutions for urban management and planning, recommending an optimal set of improvement strategies. The system should analyze all possible improvement strategies and tradeoffs, balancing required budgets and expected benefits. For example, focusing on the reference application scenario of trajectory analysis, a DSS can easy the traffic control operations allowing for the selection and best implementation of traffic management measures in response to the occurrence of an incident, with the aim to minimize incident negative consequences on traffic, as formation of congestion and queues, travel delays and risks for secondary incidents.
5. *Visualisation*: the outputs of data processing can be visualized in a user-friendly way by using various graphical user interface (GUI) tools, simulations and maps for various platforms such as web and smartphones (e.g., using google map standards).
6. *Social-networking*: users interact to share experiences and information (e.g., twitter, forums).

Due to the complexity of urban-related activities within a smart city context, we aim to provide an integrated computing environment for composing and running applications in the smart city area, leaving the user free to work at the application level and not at the middleware programming level. To this aim, one of the main objective of the framework is to assist users in formulating problems, allowing to compare different available applications (and choosing among them) to solve a given problem, or to define a new application as composition of available data and software components. The Cloud computing paradigm allows to implement the above urban-related services: facilitates data access and storage across platforms, provides on-demand computational resources, and allows for integrated processing and data analysis. The proposed architecture is based on the use of both proprietary and open source software and the integration of both private and public-available environmental databases.

Figure 1 shows the architecture consisting of a set of layers. At the lower level, the *platform layer* is based on a hybrid Cloud environment (also including mobile devices at any type) that ensures cross-platform accessibility of environmental data.

The *data acquisition layer* allows to access environmental data collected from disparate sources including remote database repositories, sensor networks, mobile ad-hoc networks, satellites, camera, meters, etc. The devices at this level provide measures and sensed values that are gathered by a set of software modules that before storing them, classify the collected data into application specific thematic categories, perform pre-processing and/or data aggregation and update the data/service catalogues for further use of the data. The activities found at this level can measure

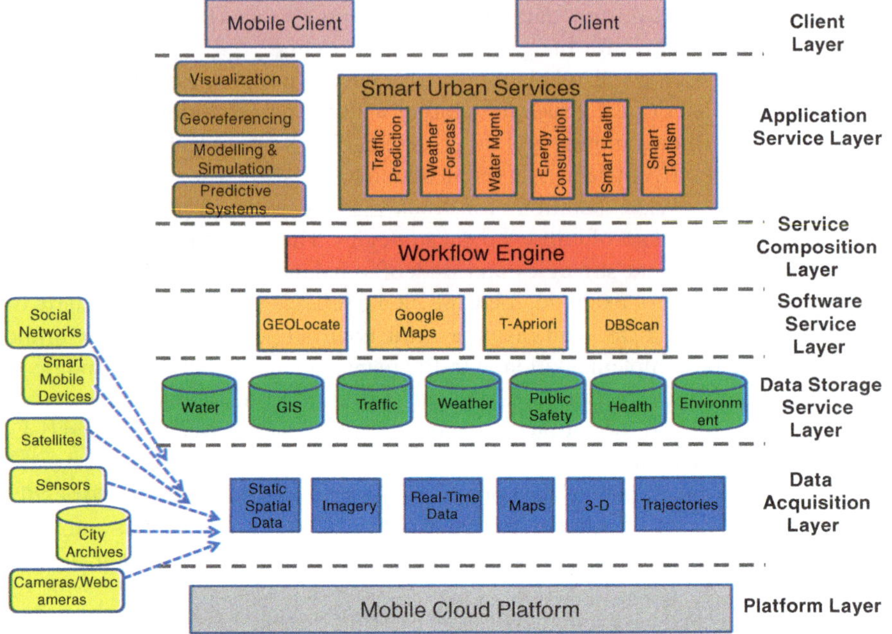

Fig. 1 A Cloud-based architecture for smart cities

water quality, collect electrical meter readings for a power grid, or provide building measurements to determine its energy usage.

At the *data storage level* the data collected is organized in ad-hoc repositories. More precisely, two types of repositories are available: those composing the historical archiving of specific urban features and real-time repositories that maintain only recently sensed data and their update-rate is high.

The *software service layer* is composed of a set of software components exposed as services. Those services use data provided by the lower level and are invoked by the upper level to compose applications. This layer includes, among the others, data mining algorithms (e.g., DBScan, T-Apriori) and open-source GIS tools (e.g., QGIS, GEOLocate, Google Maps, Google Earth, MapWinGIS, Image Georeferencer).

The *service composition layer* is responsible to design workflows, identify data sources, and link necessary processing components to enact the workflows. This layer is also responsible for the triggering of predefined workflows to handle the specific situations, whether it is an emergency response or the application for a citizen service. Further, this layer enables stakeholders to use existing tools and develop new application domain specific components and services.

The *Smart urban application services layer* offers a set of services for urban management. As the purpose of the platform is to present a high-level management view within a set of domains, it provides means to model and define application-specific performance indicators by implementing a set of software modules each of which addresses a specific urban feature. Existing applications (e.g, the outcomes from the service composition layer in application domain specific tools) can be used to perform intelligent analysis on environmental data.

4 Trajectory Data Analysis

This section provides a real-world application scenario as a case study of urban planning and management within the proposed framework. This scenario concerns trajectory data analysis by exploiting mobile user movements. In particular, we focused on the discovery of *trajectory patterns* from a dataset of trajectories, recently emerged as a challenging task.

In the remainder of the section is first briefly presented the focused application scenario (4.1). Then, it is illustrated the proposed trajectory pattern detection methodology (4.2) and its parallel implementation is introduced (4.3). In particular, it is shown how a workflow mechanism can be used to design the proposed methodology within a parallel setting as the one of the proposed Cloud-based smart city architecture.

4.1 Application Scenario: Trajectory Pattern Mining

In this section we introduce one scenario as a case study. This scenario focuses on the study of the trajectories followed by mobile devices with the aim to understand trajectories in order to catch user's behaviour and provide useful information about mobility-related phenomena. In particular, we apply a trajectory pattern extraction methodology allowing to predict future movements of citizens and exploit this information to support decisions in various ways. In particular, the trajectory patterns extracted represent a basic building block around which further analysis can be constructed, including the following ones:

- Predict the possible future location of a moving object from trajectory patterns using the objects recent movements. This allows to anticipate or pre-fetch possible services in the next location.
- Intelligent traffic management. Analysis of traffic congestion, providing real-time updates on traffic flow, allowing for predictions of driving time. Congestion patterns can be predicted days in advance, and traffic jams detected before they become serious.
- Estimate the similarity between users in terms of their location histories so as to promote, for example, car sharing, etc.
- Travel recommendations, e.g., mine the top interesting locations and travel sequences among locations. For example, giving a set of locations of interests it is

possible to determine the transportation modes that people took to get that place or for example establish the best route to follow to reach a given location.

The problem of discovering periodic patterns from historical object movements is very challenging. In [17] an approach to discovery hidden periodic patterns in spatio-temporal data is proposed. In particular, authors define the spatio-temporal periodic pattern mining problem and propose an algorithm for retrieving maximal periodic patterns. Moreover, they devise a specialized index structure, aimed at supporting more efficient execution of spatiotemporal queries over the discovered patterns.

A prediction approach to estimate an object future location, based on its pattern information and recent movements, is proposed in [16]. Specifically, the discovered trajectory patterns are stored in the TPT, a tree data structure exploited for an efficient and accurate prediction of future locations. In addition, two query processing techniques are presented, to perform both near and distant time predictive queries on the TPT structure.

Reference [24] presents a smart driving direction system, where GPS-equipped taxis are employed as mobile sensors aimed at probing the traffic rhythm of a city. In particular, the main idea is to exploit the intelligence of experienced taxi drivers so as to provide a user with the practically fastest route to a given destination at a given departure time. The system has been tested on a real-world trajectory dataset generated by over thirty thousands taxis in a period of 3 months, aimed at evaluating the effectiveness of the approach.

In [13] the authors extend the sequential pattern mining methodology to analyse moving objects. Some approaches of different complexity are proposed, that have been empirically evaluated over real data and synthetic benchmarks, comparing their strengths and weaknesses.

4.2 Trajectory Pattern Detection Methodology

This section describes the approach adopted to detect trajectory patterns, whose inspiring idea is common to other approaches proposed in literature [16, 18]. Specifically, the discovery process is composed of three steps, as drawn in Fig. 2.

1. The first step consists in the *detection of frequent regions* from the original trajectory dataset. The goal of this step is detecting spatial areas more densely passed through, in order to conduct the further analysis as movements through areas rather than single points.
2. The second step consists in the *synthesization of the trajectories*, by changing their representation from movements between points into movements between dense regions. Precisely, each point of the original dataset is substituted by the dense regions it belongs to.
3. The third step is aimed at *extracting trajectory patterns*, in the form of associative rules, analyzing the trajectories of dense regions obtained at the previous step.

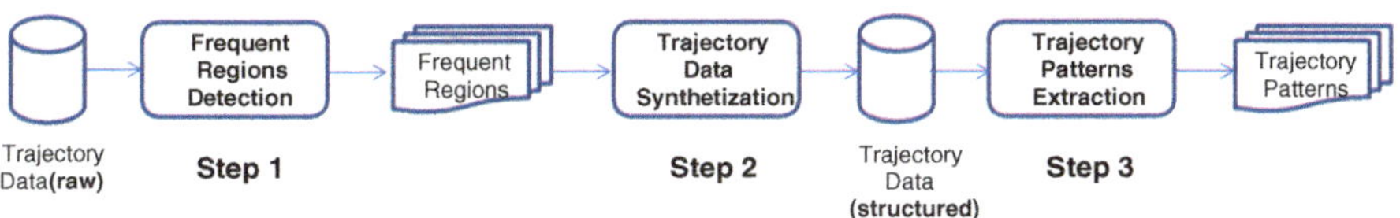

Fig. 2 Trajectory pattern detection steps

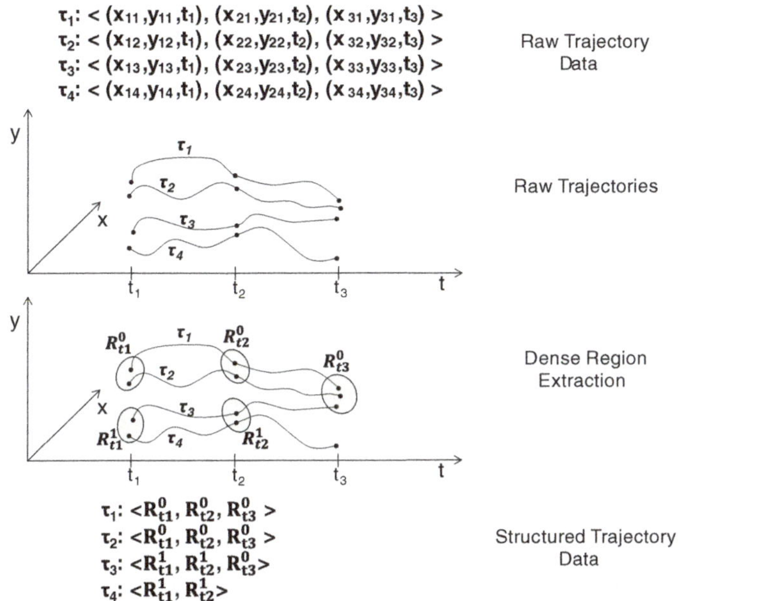

Fig. 3 Trajectory projection from raw to structured representations

In the following some notation used in the remainder of the chapter is introduced. Let be $T = < t_1, t_2, \ldots, t_H >$ a timestamp sequence, such that $t_h < t_{h+1}, \forall_{0<h<H}$. A *trajectory* τ_K is a spatio-temporal sequence, $\tau_K = < (x_{1K}, y_{1K}, t_1), \ldots, (x_{HK}, y_{HK}, t_H) >$, where each triple (x_{iK}, y_{iK}, t_i) indicates that an object of the trajectory τ_K is in the position (x_{iK}, y_{iK}) at time t_i. The *trajectory length* is the number of triples composing the trajectory (i.e., H). A *dense region* is an area of points that is more frequently visited by the object's trajectories with respect to other areas; in particular, we represent with R_t^j the jth dense region at the time t. A *structured trajectory* τ_K is a spatio-temporal sequence, $\tau_K =< R_{t_1}^{j_1}, \ldots, R_{t_H}^{j_H} >$, where each element $R_{t_i}^{j_i}$ indicates that an object of the trajectory τ_K is in the dense region R^{j_i} at time t_i (see Fig. 3). A *trajectory pattern* is a special association rule, in the form $R_{t_1}^{j_1} \wedge R_{t_2}^{j_2} \wedge \ldots \wedge R_{t_r}^{j_r} \rightarrow^c R_{t_s}^{j_s}$, with time constraints $t_1 < t_2 < \ldots < t_r < t_s$. The block on the left, i.e., $R_{t_1}^{j_1} \wedge R_{t_2}^{j_2} \wedge \ldots \wedge R_{t_r}^{j_r}$ is the premise, while $R_{t_s}^{j_s}$ is the consequence of the rule. Finally, c is the confidence of the rule, meaning that when the premise occurs then the consequence will occur with probability c.

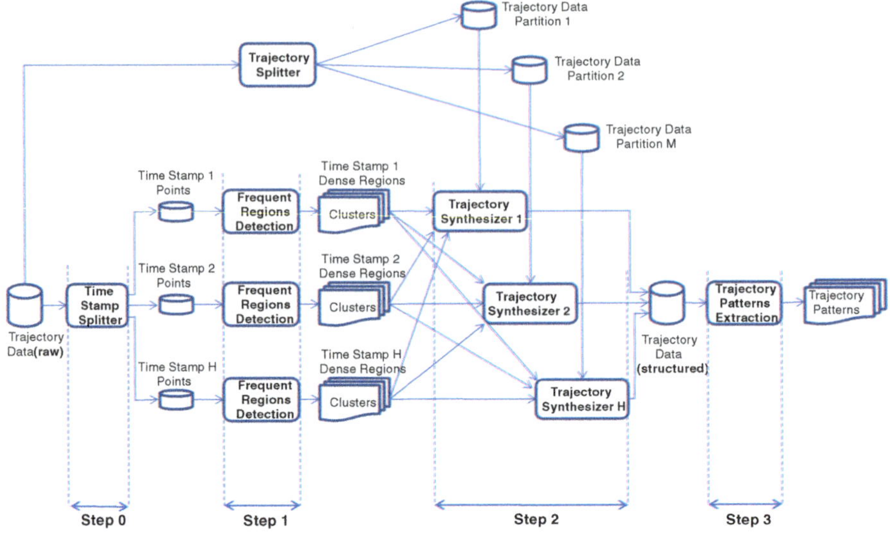

Fig. 4 Trajectory pattern detection workflow

4.3 Trajectory Pattern Detection: A Parallel Implementation

The trajectory pattern detection process consists of a sequence of concatenated connected steps involving different kinds of data and tools that can be located over geographically distributed environments. Thus, it may be suitably represented by exploiting the workflow formalism, i.e., a graph in which nodes represent data sources, data mining tools and algorithms, and edges represent execution dependencies among nodes.

The workflow representing the proposed trajectory pattern detection methodology is shown in Fig. 4. The original data set D is a raw trajectory data, populated by the trajectories (represented in the previously described format) of some users collected somehow. In particular, let us suppose that the original dataset is composed of N trajectories, each one represented as a sequence of H (x, y, t)-triples.

Before going into the details of the application workflow, we give a brief description of the tools and algorithms involved in our application:

- *Time Stamp Splitter*: it receives a trajectory dataset and performs a *vertical* partitioning on it, by storing in the ith output dataset all the points that have been covered at the ith timestamp;
- *Trajectory Splitter*: it receives a trajectory dataset and performs an *horizontal* partitioning on it, by splitting the original trajectory set in M partitions of equivalent sizes;
- *Trajectory Synthesizer*: it rebuilds the trajectories, by starting from the original dataset and taking into account the points belonging to the frequent regions. In particular, it creates a new trajectory dataset, by substituting each point with the dense

region it belongs to. The final dataset results populated by trajectories between dense regions (but between single points);

- *Frequent Regions Detection Algorithm*: it detects frequent regions from the original trajectory dataset, in order to discover geographical areas that are more densely visited with respect to others; this step can be well performed by a density-based clustering algorithm (e.g DBSCAN [21] or OPTICS [10]), whose detected clusters are defined as areas of higher density than the remainder of the data set.
- *Trajectory Pattern Extraction Algorithm*: it discovers trajectory patterns, that can be seen as concise descriptions of frequent behaviours, in terms of both space (i.e., the regions of space visited during movements) and time (i.e., the duration of movements). This step can be done by different algorithms (some approaches are presented in [16, 17]), with the goal of extract rules whose elements respect a monotonically increasing order in terms of time offset associated to each one; that is, the timestamps of the antecedents are chronologically previous of those appearing in the consequent.

The above described workflow can be implemented within the proposed Cloud-based architecture that allows for composing, compiling and running urban-related applications on the Cloud. This solution allow us to split the datasets and distribute the computation on different nodes, in order to achieve better performances. Specifically, the raw trajectories data can be gathered through the Data Acquisition Layer and then stored in a database (or over multiple distributed databases) by means of the Data Storage Level. The workflow can be designed and implemented through the Service Composition Layer that interacts with both the Data Storage Service layer and the Software Service Layer (where are located the DBScan and T-Apriori algorithms) to compose the concatenated sequence of tasks.

In the following the parallel version of the workflow is described (see Fig. 4) .

Step 0—Vertical Data Splitting. The original trajectory dataset is partitioned by the *Time Stamp Splitter* in a vertical way. In other words, all the points visited at the same time stamp $t_i \in T$ will be gathered in the same dataset. At the end of this step, $|T|$ different datasets are available.

Step 1—Frequent Regions Detection. At this point, frequent regions are detected by the parallel execution of $|T|$ density-based clustering algorithm instances, each one performed on one of the datasets built at the previous step. The result of this step consists of $|T|$ clustering models, each one corresponding to a single timestamp. Each cluster of the tth model corresponds to a dense region. The number of detected regions (i.e., number of clusters) may be different for each timestamp t.

Step 2—Trajectory Data Synthetization. The original dataset is horizontally split in M partitions, each one containing N/M trajectories (or tuples). Such datasets are represented in figure as *TrajectoryDataPartition* icons. Now, starting from these datasets and the dense regions (discovered at the previous step), the composition of the trajectory data as movements between dense regions (and not more between points) is done. This task is performed by $|T|$ instances of the *Trajectory Synthesizer* tool, each one building a new dataset of dense region trajectories, by considering all the points of the original trajectories belonging to the dense regions (discovered at

the Step 1). All these trajectories are collected in a final dataset (*Trajectory Data (structured)* in figure).
Step 3—Trajectory Pattern Extraction. Finally, a *Trajectory Pattern Extraction* algorithm on the dense regions trajectory data is executed, to discover trajectory patterns from them. The final mining model is a set of associative rules describing spatio-temporal relations between the movement of the users under investigation.

5 Experimental Evaluation

In this section, we explore a real case study by applying the trajectory pattern detection method described in this chapter over a real-world dataset. The input dataset chosen for the experiments is the *T-Drive Trajectory Data Sample* [23, 25], a collection of GPS traces describing the movement of GPS-equipped cars in the urban area of Beijing, China. The temporal span of the dataset is one week. The number of vehicles tracked is 10,357 taxis. The total number of points is about 15 millions and the total area covered by the trajectories reaches almost 9 million km. Starting from this dataset, we extracted three different subsets of 25000, 50000 and 100000 trajectories respectively, obtained by sampling taxi positions every 5 min. In particular, trajectories are composed of $|T| = 100$ samples.

To evaluate the effectiveness and the efficiency of the proposed approach, we carried out a performance analysis by executing different experiments in various scenarios. Experiments have been conducted on Intel Pentium 4 processor machines, with CPU frequency 1.36 GHz and 2 GB RAM.

The experimental evaluation aimed at analyzing the execution time of the trajectory pattern task, by evaluating the time elapsed in each step and comparing the performances obtained by both sequential and parallel implementation.

Let us start by analyzing the execution time of the sequential algorithm, whose steps are reported in Fig. 2. In particular, the three steps of the sequential flow, i.e., the Frequent Regions Detection, the Trajectory Data Synthesization and the Trajectory Pattern Extraction (i.e., Step 1, 2 and 3) have been implemented by the DBScan, the Trajectory Synthesizer and the T-Apriori tools respectively. Let T_S be the *sequential execution time*, i.e., the time needed to execute all steps sequentially. Formally, the total time spent to complete the application can be expressed as

$$T_S = T_{S1} + T_{S2} + T_{S3} \tag{1}$$

where T_{Si} represents the time spent to execute the ith step in the sequential execution.

Figure 5 shows the execution time (in a logarithmic scale) of each step with respect to different data sizes, i.e., number of trajectories. It is worth noticing that the Dense Region extraction is the most time consuming step of the whole process (taking more than 99 % of the total execution time) and the time spent by the other two steps is negligible with respect to the first one. Such time is very long because the dense

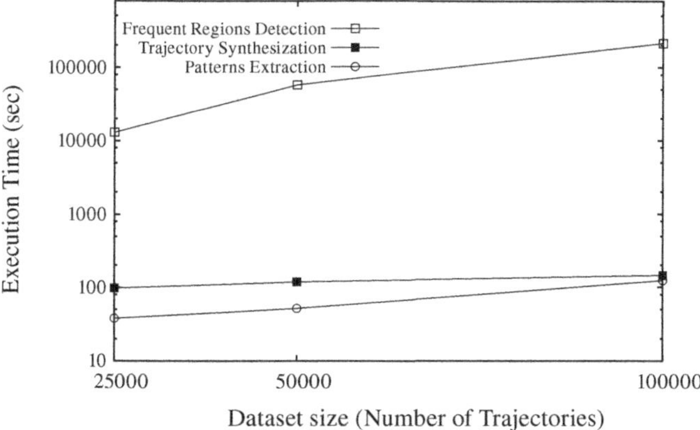

Fig. 5 Sequential scenario: execution time (logarithmic scale) of the three steps for different data size

region extraction step is performed by the execution of $|T|$ algorithm instances, each one to discover the dense regions at the tth timestamp ($t = 1 \ldots |T|$).

Now, let us observe the workflow represented in Fig. 4, where the sequential flow has been decomposed and modeled as a task-dependency graph. The *critical path* of a workflow is defined as a path with the longest average execution time from the start activity to the end activity. The amount of time needed to execute all the tasks composing the critical path is named *critical path length* [14]. Since all the paths of the workflow are executed in parallel, the critical path length determines the execution time of the whole workflow. Let T_P be the *parallel execution time*, i.e., the time needed to execute the workflow. Formally, the total time spent to complete the execution can be expressed as:

$$T_P = T_{P0} + T_{P1} + T_{P2} + T_{P3} \tag{2}$$

where T_{P0}, T_{P1}, T_{P2}, T_{P3} represents the time spent by the TimeStamp Splitter, the DBScan, the Trajectory Synthesizer and Apriori respectively (Step 1, 2, 3 and 4).

Figure 6 shows the execution time of the four steps composing the critical path of the workflow, for different data sizes, when $|T| = 100$ critical paths are executed in parallel. We can observe a considerable reduction in the execution time of the Frequent Regions Detection step compared to the sequential case. This is mainly due to a reduction of the DBSCAN computation time. In fact, the DBSCAN instance running in the critical path has to process a dataset whose size is $|D|/|T|$, resulting in a computation time that is almost $|T|$ times shorter than the sequential scenario.

This improvements can be better appreciated by observing Fig. 7, in which sequential and parallel execution times are plotted for different data sizes and different number of nodes. Parallel times are obtained considering a parallelism degree of $n = 25$, 50 and 100 nodes, respectively. It is worth noticing that the parallel execution times are strongly reduced with respect to the sequential one.

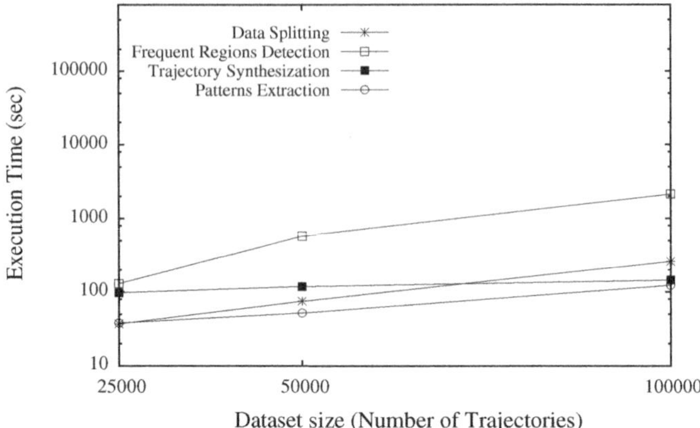

Fig. 6 Parallel scenario: execution time (logarithmic scale) of the four steps for different data size

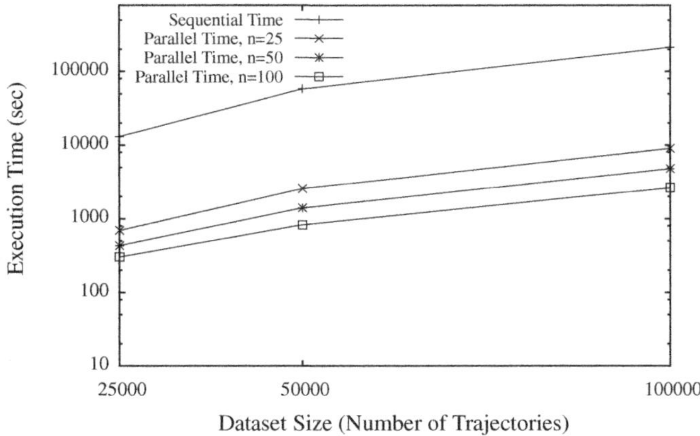

Fig. 7 Sequential and parallel execution time, for different data sizes and number of nodes

Table 1 Parallel scenario: execution time and relative speedup versus the number of available nodes, for $|D| = 100\,\mathrm{K}$

Number of nodes	Execution time (s)	Speedup
1	213923	1
25	9000	23.7
50	4746	45.0
100	2668	80.1

Table 1 reports the total execution time and speedup with respect to different numbers of available nodes, for $|D| = 100\,\mathrm{K}$ trajectories. The speedup is the ratio of the turnaround time obtained with a single node to the turnaround time computed with n nodes. In particular, Fig. 8 plots the achieved execution speedup: it ranges from 23.7 (using 25 nodes) to 80.1 (using 100 nodes). Such experiments show encouraging trends that can be studied by a further and more detailed analysis.

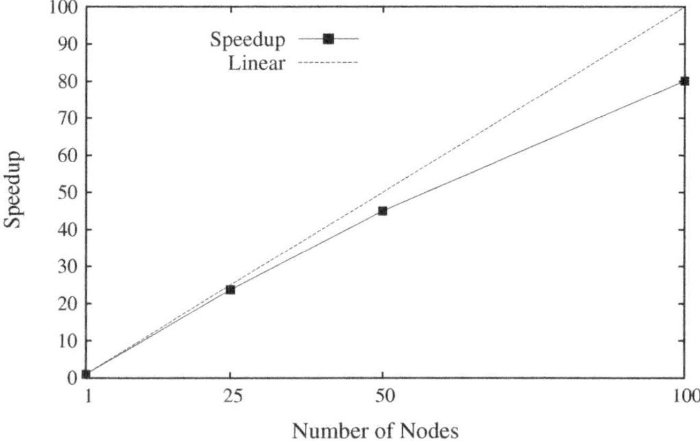

Fig. 8 Speedup versus the number of available nodes

6 Conclusions

A large amount of movement data is daily collected, due to the increasing pervasiveness of mobile and wireless devices, sensing technologies, GPS traces, etc. Such collection of information can be analyzed to discover descriptive and predictive models, that can be exploited to have a smart management of the city resources and to generally improve urban life.

This chapter presents an architecture of a Cloud-based framework for urban computing, aimed at supporting smart cities development. Within such framework we have described a parallel approach, modeled by the workflow formalism, for pattern discovery from trajectory data. The main idea consists in *(i)* finding the more densely passed through regions in a given geographical area, and *(ii)* then extracting trajectory patterns from those regions in the form of association rules. Experimental evaluation, conducted on a real-world dataset, shows that the trajectory pattern mining process takes advantage from a parallel execution environment as offered by a Cloud architecture.

As future work, our research will proceed in different directions. First, we will introduce some optimizations on the trajectory analysis methodology proposed in the chapter, to improve the efficiency and effectiveness of the approach. Second, we are developing a visualization module for the input data and discovered models.

Acknowledgments This research work has been partially funded by the MIUR projects TETRIS (PON01_00451) and DICET-INMOTO (PON04a2_D).

References

1. http://ec.europa.eu/information/society/activities/livinglabs/docs/epi/cv6/pub.pdf
2. http://public.dhe.ibm.com/common/ssi/ecm/en/giw03021usen/giw03021usen.pdf
3. http://puppetlabs.com/puppet/puppet-enterprise
4. http://www.environmentalvirtualobservatory.org/
5. http://www.ess.co.at/
6. http://www.life2project.eu
7. http://www.nist.gov/itl/cloud/cloud052009.cfm
8. http://www.nivis.com/technology/smartobject.php
9. http://www.openqrm-enterprise.com/
10. Ankerst, M., Breunig, M.M., peter Kriegel, H., Sander, J.: Optics: ordering points to identify the clustering structure. In: Proceedings of the 1999 ACM SIGMOD International Conference on Management of Data, SIGMOD '99. ACM, pp. 49–60 (1999)
11. Anthopoulos, L.G., Vakali, A.: The future internet, chap. Urban Planning and Smart Cities: Interrelations and Reciprocities. Springer, Berlin, pp. 178–189 (2012)
12. Cesario, E., Comito, C., Talia, D.: Towards a cloud-based framework for urban computing. The trajectory analysis case. In: Proceedings of the 3rd International Conference on Cloud and Green Computing, CGC '13. IEEE (2013). To appear
13. Giannotti, F., Nanni, M., Pinelli, F., Pedreschi, D.: Trajectory pattern mining. In: Proceedings of the 13th ACM SIGKDD International Conference on Knowledge Discovery and Data Mining, KDD '07. ACM, pp. 330–339 (2007)
14. Grama, A., Karypis, G., Kumar, V., Gupta, A.: Introduction to Parallel Computing. Pearson (2003)
15. Jackson, K.: OpenStack Cloud Computing Cookbook. Packt Publishing (2012)
16. Jeung, H., Liu, Q., Shen, H., Tao Zhou, X.: A hybrid prediction model for moving objects. In: Proceedings of the 2008 IEEE 24th International Conference on Data Engineering, ICDE '08, IEEE Computer Society, pp. 70–79 (2008)
17. Mamoulis, N., Cao, H., Kollios, G., Hadjieleftheriou, M., Tao, Y., Cheung, D.W.: Mining, indexing, and querying historical spatiotemporal data. In: Proceedings of the Tenth ACM SIGKDD International Conference on Knowledge Discovery and Data Mining, KDD '04. ACM, pp. 236–245 (2004)
18. Monreale, A., Pinelli, F., Trasarti, R., Giannotti, F.: Wherenext: a location predictor on trajectory pattern mining. In: Proceedings of the 15th ACM SIGKDD International Conference on Knowledge Discovery and Data Mining, KDD '09. ACM, pp. 637–646 (2009)
19. Moreno-Vozmediano, R., Montero, R.S., Llorente, I.M.: Iaas cloud architecture: From virtualized datacenters to federated cloud infrastructures. Computer **45**(12), 65–72 (2012)
20. Nurmi, D., Wolski, R., Grzegorczyk, C., Obertelli, G., Soman, S., Youseff, L., Zagorodnov, D.: The eucalyptus open-source cloud-computing system. In: Proceedings of the 2009 9th IEEE/ACM International Symposium on Cluster Computing and the Grid, CCGRID '09. IEEE Computer Society, pp. 124–131 (2009)
21. Sander, J., Ester, M., Kriegel, H.P., Xu, X.: Density-based clustering in spatial databases: the algorithm gdbscan and its applications. Data Min. Knowl. Discov. **2**(2), 169–194 (1998)
22. Talia, D.: Clouds for scalable big data analytics. Computer **46**(5), 98–101 (2013)
23. Yuan, J., Zheng, Y., Xie, X., Sun, G.: Driving with knowledge from the physical world. In: Proceedings of the 17th ACM SIGKDD International Conference on Knowledge Discovery and Data Mining, KDD '11. ACM, pp. 316–324 (2011)
24. Yuan, J., Zheng, Y., Xie, X., Sun, G.: T-drive: enhancing driving directions with taxi drivers' intelligence. IEEE Trans. Knowl. Data Eng. **25**(1), 220–232 (2013)
25. Yuan, J., Zheng, Y., Zhang, C., Xie, W., Xie, X., Sun, G., Huang, Y.: T-drive: driving directions based on taxi trajectories. In: Proceedings of the 18th SIGSPATIAL International Conference on Advances in Geographic Information Systems, GIS '10. ACM, pp. 99–108 (2010)

People-Centric Service for mHealth of Wheelchair Users in Smart Cities

Lin Yang, Wenfeng Li, Yanhong Ge, Xiuwen Fu, Raffaele Gravina and Giancarlo Fortino

Abstract Urban dwellers are soul of smart cities, and all final aims of city applications are people-centric. mHealth is a new generation method for personal healthcare, specially smart phone is widely used to interact with surroundings by the disabled and elderly people in smart cities. Existing massive of sensors, actuators, and smart objects are separated and controlled in different owners and community. Mobile devices of people-centric sensing (PCS) can receive data in opportunistic sensing according to mobile geo-location, dynamic social relationship, and interests of people, etc. In this work, we present a real-time health-driven model for people-centric healthcare context, and present a social-aware architecture to support smart objects mapping to online social networks, then present discovering and interacting with shared smart objects in a virtual community. Finally, we present a prototype system to validate the people-centric mHealth service model.

Keywords Smart cities · People-centric · mHealth · Smart object · Mobile social wireless sensor networks

L. Yang · W. Li (✉) · Y. Ge · X. Fu
Wuhan University of Technology, Wuhan, China
e-mail: liwf@whut.edu.cn

L. Yang
e-mail: Lyang@whut.edu.cn

R. Gravina
University of Calabria, Via Pietro Bucci, 87036 Rende, CS, Italy
e-mail: rgravina@unical.it

G. Fortino
DIMES, University of Calabria, Rende (CS), Italy
e-mail: g.fortino@unical.it

G. Fortino and P. Trunfio (eds.), *Internet of Things Based on Smart Objects*,
Internet of Things, DOI: 10.1007/978-3-319-00491-4_9,

1 Introduction

mHealth is an efficient method for healthcare monitoring application of mobile and large-scale daily activities [1]. Development and treatment of chronic diseases take place in daily life outside of traditional clinical settings. To determine and adjust treatment for these diseases, clinicians depend heavily on patient reports of symptoms, side effects, and functional status. In particular, for the problem of increasing elderly and disabled people living alone, it's necessary to develop new technology for real-time monitoring the health status and interacting with living environment.

People-centric sensing network is suitable for large-scale, mobile and opportunistic perception application in urban area, and obtaining individual information from the big data of cities via the people's interest, location, and mission [2]. The ability of Dwellers to interact with city life environment is enhanced by using the Internet of Things (IoT). Specifically, wheelchair for elderly and disabled people, integrating the new generation information technology of smart object, social mobile computing, is a convenient tool for monitoring real-time healthcare and interacting with environment according to the real-time health status, while they are living alone.

Community is defined to identify local device/node's membership. People interact with different physical device community to transform data and information via social-driven methods in daily life. Existing IoT architecture supports physical sensors or actuators virtualized to virtual nodes or IP-based resources on the internet; physical objects are mapped to a cyber-world. The online resource is easy to be inquired by mobile application [3–6]. Virtual nodes form a virtual community according to function, physical communication range, physical topology and the device ownership etc. Wireless body area networks, smart wheelchair and smart phone build a people-centric physical community to complete a daily life healthcare monitoring task. And a mobile people-centric healthcare device gets the shared data from the large-scale environment in smart cities.

There are some issues in data sharing in smart cities. First, in public place, different organizers build their own devices to complete the same sensing mission, and that wastes money. Second, smart objects and surrounding environment are published to the internet by different owners, while a smart wheelchair moves to unfamiliar room or buildings, the wheelchair user need to get access to the shared resources by social activity with the owners.

For a mobility of wheelchair user activities, we present a novel online mobile social network to improve the efficiency of resource sharing. Section 2 presents a context model of the real-time health status-driven model for people-centric sensing. Section 3 presents an architecture of wheelchair users' HealthCare device and social aware method. Section 4 presents prototype system implement and discussion, and Sect. 5 presents conclusion and future work.

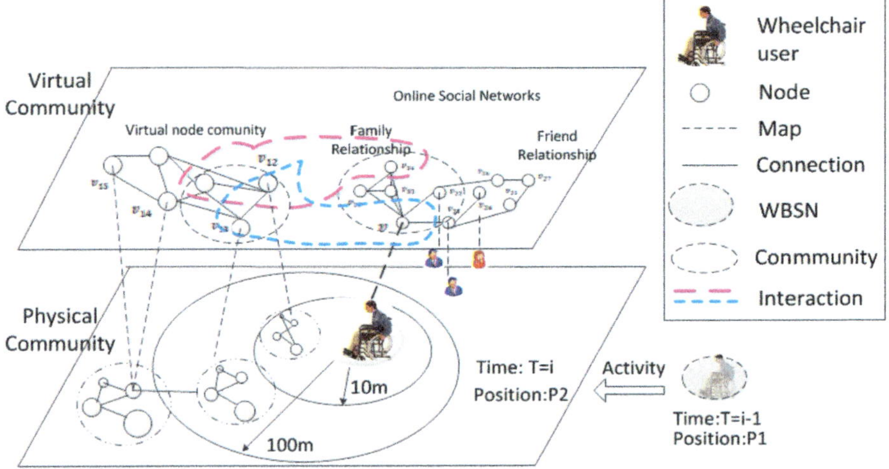

Fig. 1 Community of human, physical and virtual nodes

2 Model of People-Centric Healthcare for the Disabled

As shown in Fig. 1, a wheelchair user moves around a city, and that leads to opportunistic data exchanging in physical world and virtual social world. In physical community, opportunistic communications are created to interact with surrounding environments via the mobility of human or carried devices, such as smart phone, wheelchair and vehicles. The social activity of person brings new human relationship in social community. As a virtual node, the person is added into online social networks to exchange their own shared devices, and that is aim to get access to unknown devices unconnected directly before. Especially smart phone and smart object technologies improve the people-centric healthcare application, as smart objects can be used to detect vital physiological parameters and requirements. Compared to the existing context-driven social sensing and analyzing, the people-centric healthcare perception realize the real-time requirement more truthful, accurate, and the physical status-driven model can get accurate recommendation from big data. Section 2.1 presents context modeling of people-centric healthcare. And Sect. 2.2 presents resource discovering in unknown environment.

2.1 Context Modelling

Context models can be compared in terms of efficiency (to access data and execute reasoning procedures), scalability, and usability of the formalism [7]. Figure 2 presents people-centric healthcare context modeling, and depicts four parts: device, health-context, social aware, and location.

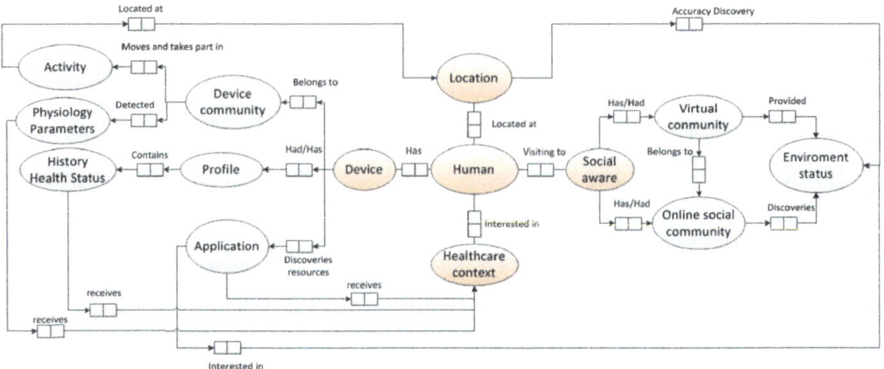

Fig. 2 People-centric healthcare context model

- *Device*—Human has kinds of devices, such as smart phone or a wheelchair. And the devices store and update the fixed attribute of owner and devices, such as chronic disease, historical health status, social information, the information of device itself etc. and device is not only the tool for perception of local information, but also the interface for interacting with cyber-physical world. For example, wheelchair can perceive user's health status in real-time as well as the surrounding environment. The user, who carries the device as a whole people aware system, moves into a community of other physical devices for exchanging data. While people take the wheelchair into the different buildings, across different wireless sensor networks, wireless sensor network can obtain data from different buildings. People participate in social activities for having chance to get the new physical location and social relations.
- *Location*—An activity of device located at location information. The mobility of device creates dynamic physical position and online social community identification. The location is the trigger to accuracy discovering surrounding resources from online social community. The mobile wheelchair adds to the unknown physical device communication network or get the access device belongs to other person.
- *Healthcare context*—the healthcare context model describes the rules of interoperation of online social resource, virtual history health information and real-time healthcare diagnosis from the WBSN and smart wheelchair. The individual rule comes and updates from device module via dynamic interacting with the communities. For example, physical device was visited by different person in different time in cities, and the virtual nodes was ranked priority and weighted in virtual community. Compared to the existing context definition,, healthcare context model is from user participation and real-time diagnosis of physical device. Life critical physiological warning thresholds can be input via smart phone, and priority of the emergency contact list can be ranked by records of recent visited friends.
- *Social aware*—Social aware contains virtual community of resource and online social networks. Human has or had the community via social activity to build or

destroy the connection among human, physical device, social networks. IP-based smart object is mapped to an independent service, and is virtualized as a virtual node. The dynamic statuses of virtual nodes change the division of virtual community. When a physical device is sold, rent or broken, the virtual node quiets the current communication. Surrounding environment status is accuracy discovered by trigger of "location", and captured by application of mobile devices.

2.2 Online Social Resource and Community Discovery

In mobile daily life, by using GPS device, RFID tags and WSN to geographically locate the position of wheelchair, physical environment status and social networks are discovered directly or indirectly, while healthcare device detects real-time abnormal physiological parameters or requirement of interacting with surrounding environment.

Step 1: Input the location P, accuracy discovery community of virtual nodes in the position registered.

Step 2: Inquiry community, get the resource community V of the best correlation with healthcare context, according to priority of communities' weight in community, the geo-range of search radius, depth of the community traversal. Then wheelchair connects the found community, gets the resource and labeled the communities' weight +1. Return access community.

Step 3: Get the resource of virtual nodes. If the found community is shared nodes in public, get the resource N, and label weight of nodes +1. Else return the owner or manager of the community of virtual nodes, and then inquire in the online social networks.

Step 4: Inquire and request the access from the owner. If connected with the owner, wheelchair user connects to the online social network, label the weight +1 of visited owner, and gets access community, return to **Step 2**.

Step 5: If wheelchair user has no connected relationship with owner initialization, search the available rote path or bridges to build the connection with the own on existing online social networks tools, such as third party platforms like Facebook, Twitter, or WeiChat. Then return to **Step 4**.

Step 6: Move and create opportunistic communication. If unviable to contact with owner, visit to neighbor nodes according to the priority of weight of nodes in virtual community, or move to next geo-position in the recommendation of online social networks. Then update location P, return to **Step 1**.

3 Architectures to Support Wheelchair Social Sensing

In smart cities, activities of person bring opportunistic sensing and service, and can enhance the quality of life for wheelchair users. For the healthcare and assistance of wheelchair users in daily life, we design a people-centric healthcare system.

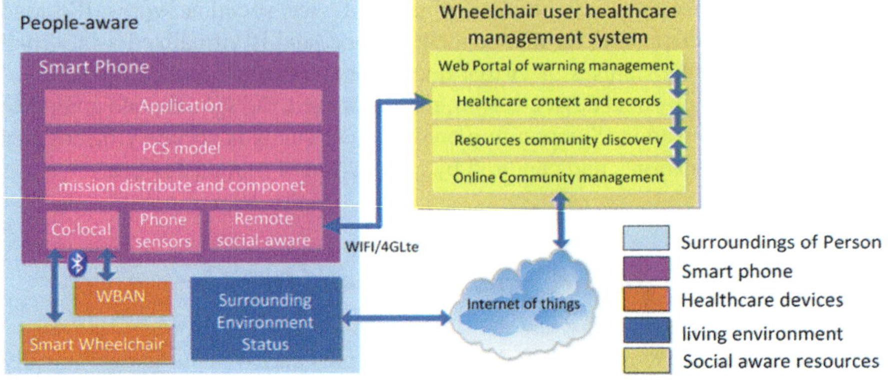

Fig. 3 Architecture of people-centric healthcare for wheelchair users

Wheelchair users get local vital physiological parameters and online shared environment status. Smart wheelchair is not only the vehicle assistant for the elderly and disabled person, but also can detect e.g. abnormal sitting status, falls of the user or the vehicle itself, and sense the surrounding environments. Sensors, actuators, and IP-enabled smart objects in the cities can be mapped in to virtual resources by using IoT technologies. The architecture is shown in Fig. 3.

3.1 Physical Device of Healthcare

(1) *Smartphone*

Smart phone is the most popular embedded device in people's daily life. The proliferation of smartphone with sensing capabilities has created an opportunity to design systems that capture vast amounts of information about people's social behavior [8]. Existing smart phones, such as Android, IPhone, Blackberry, have common features: build-in GPS module, build-in Bluetooth radio module. In this work, the GPS module is used to perceive physical location of people. And Bluetooth module is used to connect co-local smart objects, such as smart wheelchair and wireless body sensor networks. Built-in Wi-Fi or 4GLte expands the perception range via visiting Cloud sensing infrastructure.

(2) *Smart wheelchair*

Comparing with an basic wheelchair, smart wheelchair expands more functions via installing sensors and actuators. In this work, the function of smart wheelchair contains (a) user localization, (b) detecting falling from chair, (c) measuring falling of chair itself and (d) local alarm triggering.

- *RFID identification*—RFID technology is suitable for tracking goods in wireless and mobile applications [9]. A RFID reader is fixed on the wheelchair, recognizes tags carried by the user, reads the social attributions of the wheelchair user, and joins the community of human relationship. While wheelchair user move to and interact with living environments, reader in wheelchair can identify the real-time location indirectly via reading RFID tags of register goods.
- *Falling perception*—an accelerometer sensor node is fixed on wheelchair, and detects the falling status of wheelchair. Several force sensors build in the cushion of wheelchair, and detect the sitting status of person.
- *Alarm device*—buzzers and LEDs are installed on the wheelchair; while the user is detected as falling or with abnormal physiological parameters, the alarm devices are activated, and also accessed/controlled remotely, if necessary.

(3) *WBSN of Vital physiological parameter perception*

WBSN is suitable for real-time healthcare application in daily life, and dedicated frameworks offers effective programming and flexible management [10–12]. Heart rate and ECG are common indicators of health status. We present a BSN to detect the abnormal vital physiological parameters, the sensors are connected to smart phone via Bluetooth..

There are features in the smart object of people-centric aware as following: (a) wireless. Sensor nodes, actuators, and smart objects are connected with smart gateway/phone by using the IEEE 802.15.4 protocols, so there are great advantages in energy efficiency. (b) Easy to use. These non-invasive devices work continuously, without assistant need to be worn. (c) low cost. Extending sensor nodes add or remove according to individual requirement, and as a core device smartphone becomes more cheaply and widely sued in daily life.

3.2 Bring Smart Object and WSN to the Web

(1) *Smart object virtualizing*

How to integrate new generation IP-enabled smart objects is presented in this paragraph. WSN has applied on almost every urban application domains, such as localization, environment sensing, smart parking, smart light management, and intelligence transportation systems [13]. Wireless sensor networks with 6LoWPAN and IPv6 are used to perceive surrounding environment and bring objects to internet. As shown in Fig. 4, a classic intelligent node contains sensors/actuators, CPU module, radio module, battery management module. CPU module has feature of limited storage and processing ability. The logic function of CPU contains device interface and access layer, data encoder and store layer, device control and link table, message or light-weight RESTful web service. Device interface and access layer is used to transformation of A/D and encrypt/decrypt signal. Data encoder is used to provide online/off line data formalism function. The micro format of smart object or nodes

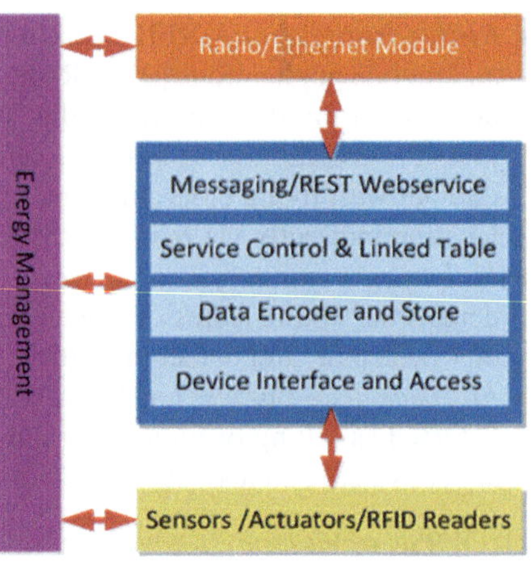

Fig. 4 Structure of a physical node

can be inquired by search engine directly, such as Geo, hCard, hProduct and hReview etc. After sematic encoding, the collected data is storage in light-weight database system, such as SQLite. The layer of service control and link table is used to define the service customer, authority, mechanism of even trigger, QoS of link. The link table layer is used to store local topology, the current rote. The physical sensor/actuator node is packaged as a virtual resource via RESTful web service, and is accessed via API.

Physical devices are packaged into service visited by short URL. Smart objects are connected to IoT platform directly or indirectly: (a) Connect directly. In future most of microcontroller support TCP/IP protocol in IP-enabled embedded device, such as 6LowPAN, IEEE 802.16.4g. (b) Smart gateway bridge nodes and IoT platform. Sensor and actuator nodes can communicate using wireless protocols such as Zigbee, Bluetooth, and ANT, or wire communications such as FiberBUS, or CAN. However, such physical nodes cannot perform data transfer via HTTP. Light weight web service is built in the smart gateway, and is aim to virtualize physical device to virtual nodes, virtual nodes are visited by URL [14]. The real communication between physical nodes and IoT platform is hidden by RESTful web service. It's easy to define a node community by the IP address, and the nodes from the same sensors/actuators networks.

(2) *Social aware resource*

Figure 5 presents software architecture of community of social resource. Resource of community can be divided into three parts: (a) data access layer, (b) cyber-physical transformation layer, and (c) sharing layer.

- *Data access layer*—Smart gateway updates data to community platform in format of JSON. RESTful Crawler collects data from the lightweight RESTful web service on internet via IPv6 protocol. We also present the following IPv4-based modules: WS-* API, SOAP web service, socket service for collecting heterogeneous resource, such as video, stream media.

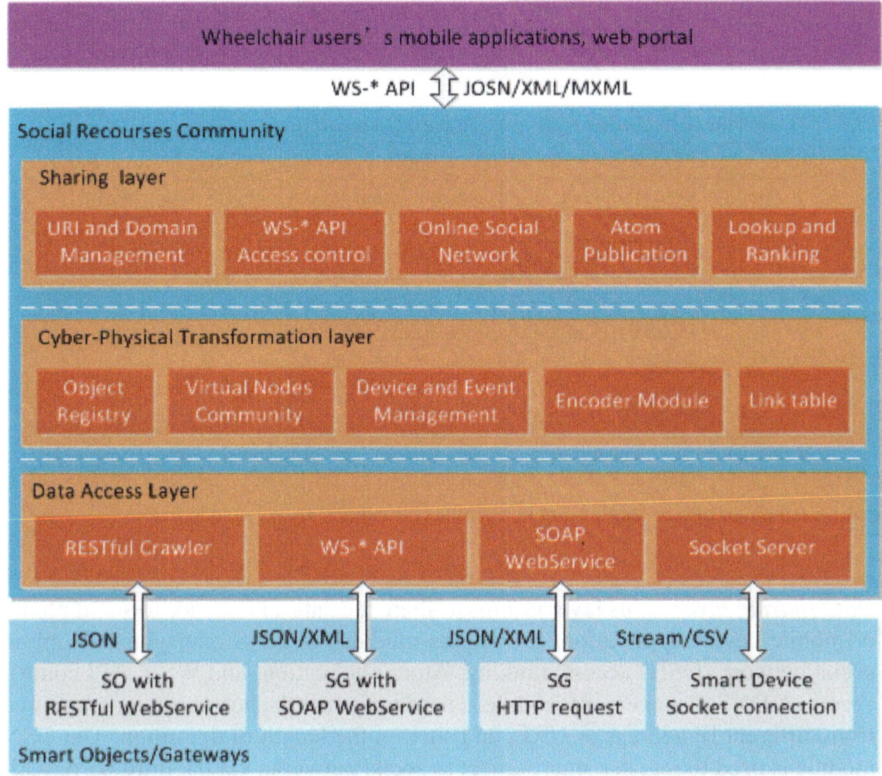

Fig. 5 Software architecture of social recourses community

- *Cyber-physical community layer*—this layer is aims to divide community and build context of formalism from the aggregated heterogeneous resource. It contains modules of object register, virtual community, device and event management, encoder module and real-time link table. Register module is used to input the initialization attribution and social ownership of physical device. Virtual community module is used to manage the division of nodes community. Module of device and event is used to record the mobility of wheelchair, social activities of person and the rule of new community division because of activities. For example, while smart wheelchair detects the abnormal vital physiological parameters, the device and event module can give recommendation of sending alarm to whom and how to, according to attributions of device and its community. Module of encoder is used to build the internal correlation via EPCglobal codes of Auto-ID. Module of link table is used to store the real-time connection relationship in the virtual nodes community. Such as neighbor of nodes, connection of communities, friendship of online social networks etc.

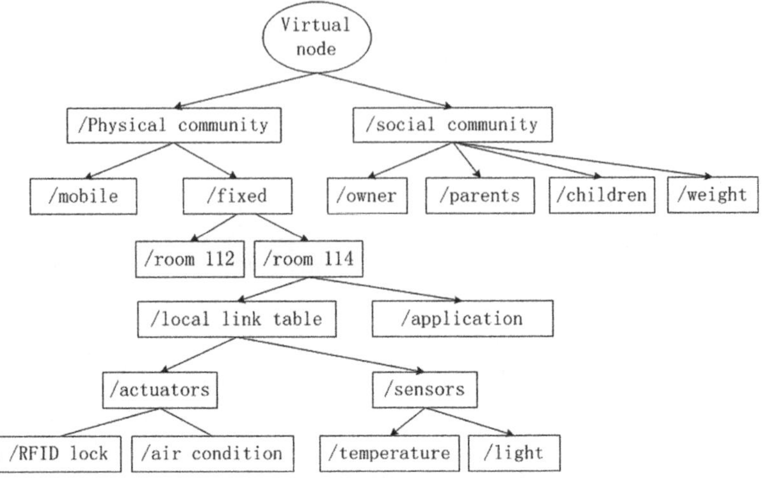

Fig. 6 A sample of device relationship in a virtual community, a virtual node indoor

- *Data sharing layer*—this layer is aims to share the data and service to smart phone or mobile device application. It contains modules of DNS management, online social network (OSN) access, ranking, Atom publication, and WS-* API control access. A DNS management module is used to format the heterogeneous resource from different IP address or URL, and shorten the length of dominion. The OSN module is used to register smart object to social networks via the third party API, such as Facebook, Twitter, or WeiChat. Each object or community is mapped to a unique ID on OSN. Thus, any resource or smart object can interact with users on OSN, such as visit, comments, link and rank. The ranking module is used to record the global ranking of visited virtual nodes and its community, and it's updated by smart phone or mobile device of wheelchair user.

3.3 Community Register

(1) *Smart wheelchair and smartphone register*

The manager or family manages the register of wheelchair and smartphone. As shown in Fig. 6, before using the devices, it is needed to input initialization information for building communities. Examples of wheelchair attributes are wheelchair user's history healthcare information, smart wheelchair sensors and actuators integrated with the social relationship of its user, address, emergency contact and online social network accounts.

(2) *Environment perceiving device register*

The owner manages the register, sharing and lifetime of environment perceiving devices. Before using devices, it's needed to input initialization information of relationship of community and sharing context.

3.4 Web Portal for Wheelchair Warning Management

Wheelchair healthcare in large-scale application necessarily requires city manager and organizations to deal with the emergency status. We present a web portal to real-time remote monitor wheelchair. Relatives can visualize the disabled person's location, vital physiological parameters and status of alarm. Figure 7 depicts a screenshot of the web portal, showing an alarm popup as a smart wheelchair detected abnormal user' health status. The manager can retrieve the alarm, the most available emergency contact from social community, and visit the surrounding environments via web portal.

4 Prototype System Implementation and Analysis

In this scenario, a wheelchair user moves into a room of his friend, and has no ownership of indoor physical objects. The aim is to get access of local shared environment status via internet, and open an RFID lock authorized by owner via OSN, while the smart wheelchair detects abnormal parameter of heart rate.

The physical devices are presented in Fig. 8. Smart wheelchair contains basic wheelchair and kinds of sensors and actuators, such as RFID reader, accelerator sensor, and force sensor in cushion etc. The human wears watch-like heart rate sensor node at wrist, ECG sensors at waist and sink node at the belt. Nodes connected to smart phone via Bluetooth module of cc2540. We install smoke, IR, humidity, temperature, light, air pollution sensors, and cameras with wireless relays nodes.. The environment gateway, based on ARM Cotex-9 and Android OS, collects data of sensors and actuators via the CC2530 Zigbee-compliant module.

As shown in Fig. 9, the proposed people-centric mHealth framework can provide access and control also in unknown environment within the urban boundaries. The result is shown on a smartphone, the platform of Android 4.04, Cortex A9 1.2 GHz, build-in Bluetooth V2.0.

As shown in Fig. 10, Smart wheelchair and WBAN can efficiently perceive the health status and surrounding environment in method of social mobile sensing. Table 1 lists measuring parameters of people-centric aware devices, and there is a little response delay on vital physiological parameters device, smart object controlling and emergency alarming. There are two factors leading to this phenomenon: (a) the mobility and activity of human body leads to near/far-field effect and shadow

Fig. 7 Web portal for wheelchair user's healthcare remote management. **a** Smart wheelchair detects abnormal physiological parameters and triggers an alarm. **b** Find the priority emergency contact person via the social community. **c** Get shared camera stream via social community discovery

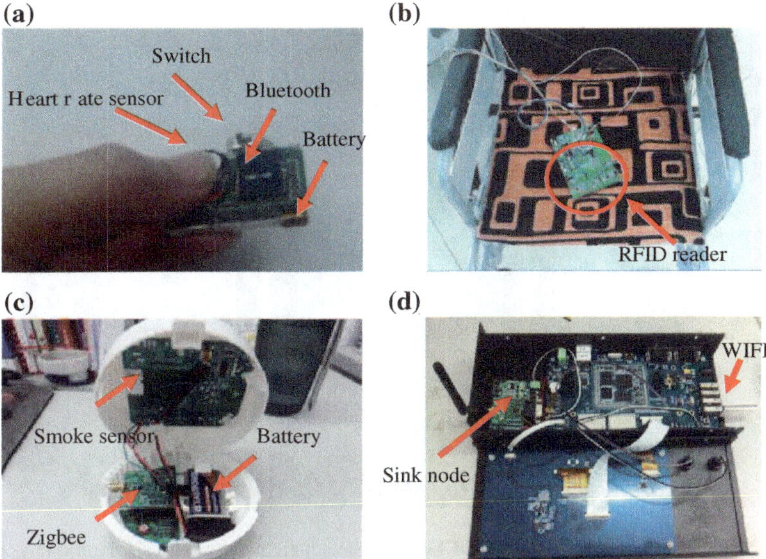

Fig. 8 Indoor sensors, smart gateway and smart wheelchair. **a** HR sensor node from WBAB devices. **b** Wheelchair integrated force cushion and RFID reader. **c** Smoke sensor from WSN of environment perception devices, **d** Smart gateway bridge Zigbee nodes to internet

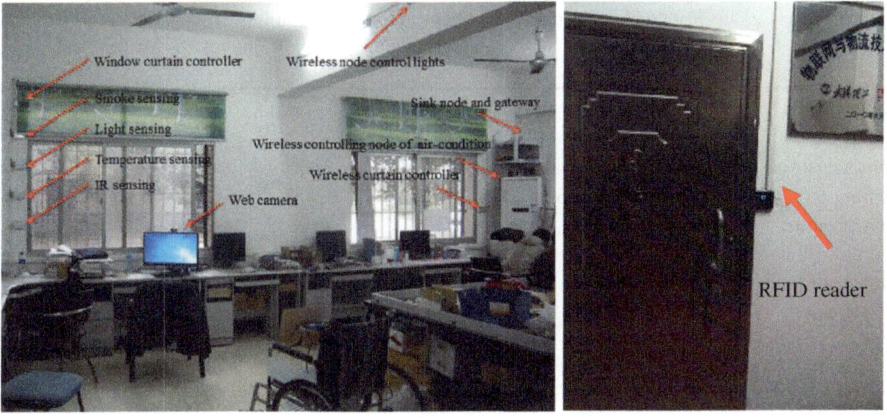

Fig. 9 Indoor WSN and RFID reader

effects, and effect the transformation of WBAN. (b) In the algorithm of detecting abnormal data of Heart Rate, ECG, falling of wheelchair, we need to select a group of continuum data to discovery features via sliding time window and comparing several features, this processing leads to time delay.

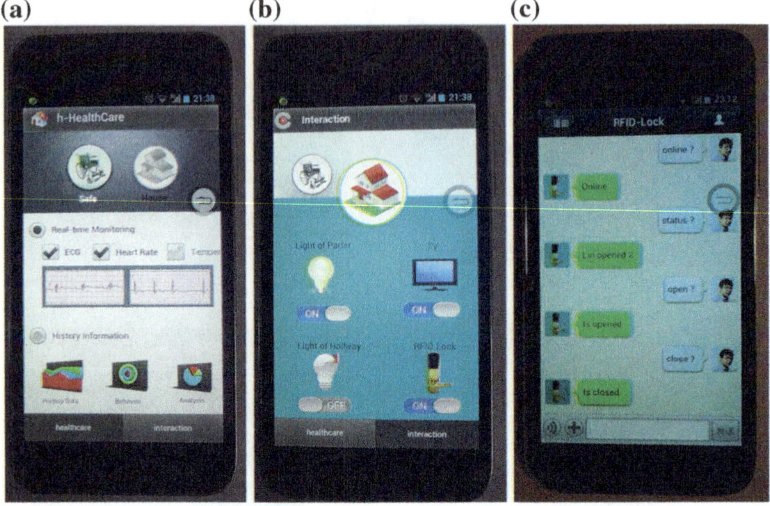

Fig. 10 Screenshot of smart phone used by the wheelchair user. **a** The smart phone perceives the vital physiology parameter of wheelchair user. **b** The smart phone controls the local shared household equipment via remote online resource. **c** A wheelchair user interacts with RFID lock via WeiChat, a IM of online social network

Table 1 Measuring parameters of healthcare devices

Parameter	Range measurement	Measurement precision	Sampling frequency (Hz)	Response time (s)
Heart rate	$0 \sim 350/\text{min}$	±2 % full	100	<2
ECG	N/A	N/A	512	<2
Smart cushion	$0.1 \sim 10\,\text{Kg/cm}^2$	0.5 % full	300	<1
Falling	$0 \sim 350\,\text{Hz}, \pm 6$ g	200 mV/g	300	<1
Temperature	$-20 \sim +80\,°\text{C}$	±0.2 °C	300	<1
Humidity	$0 \sim 100$ % RH	±2 % RH	300	<1
Household smart device	N/A	N/A	N/A	<1

5 Related Work

Existing daily healthcare systems, such as Independent LifeStyle Assistant, Aware-Home, University of Rochester's Smart Medical Home, assume the user owns the smart devices and sensors [15]. However, a wheelchair user has some social activity and enters into unknown areas, where the devices and environment sensing resources are owned and managed by others. For example, when the wheelchair moves for shopping, visiting friends, etc., it cannot interact with the surrounding environment directly. The smart wheelchair needs to use shared devices or sensors on the IoT.

For instance, OpenHealth [1] uses mobile applications and a large number of users to conduct self-health monitoring. Comparing with OpenHealth' approach, dynamic physiological parameter from the real-time smart device is more accurate; alarm triggering and resource requirement are driven by real-time health status.

The concept of smart wheelchair came into play since early 90's; an interesting work [16] surveys most of the older works. Furthermore, Simpson et al. in [17] analyze the number of people actually requiring a wheelchair, and specifically of what kind of features based on the type of motor and neurological disabilities.

It is interesting to note a lack of current literature on smart wheelchairs with integrated support for smart cities and Cloud computing technologies. This sounds peculiar as Smart City is one of the hottest topics today, along with Social Innovation project calls and the wide interests to more generic personal mobility systems, particularly in urban environments. To our knowledge, on 2007 Satoh and Sakaue [18] are the firsts to suggest the use of cellular connectivity to provide remote monitoring and support to the wheelchair user. However, only the year later, a work introducing an information server to support safety and effectiveness of urban navigation, to store status (actions, locations,...) information of the wheelchair, and to provide information to the user, has been proposed [19].

Current smart wheelchair projects are indeed mostly focused on standalone wheelchairs instrumented with sensors and specific human-machine command/feedback interfaces to add some sort of "smartness", e.g. to enable for assisted navigation support (including straight line following, obstacle/collision avoidance, and door passage) [20–22]; just in few cases authors propose user's posture/gesture detection [18, 23] or health status monitoring [23].

6 Conclusions and Future Work

In summary, this work has following characteristics: (i) smart devices get accurate "small data" from big data of cities via the real-time health status-driven; (ii) smart device can get environment status at unknown areas via online social aware; (iii) using the experienced third party platform of online social networks (OSN), smart object is brought to OSN, and operated by human in different community. Comparing with existing geo-driven or interest-driven context, this work expands the range of support services for wheelchair users, and provides more accurate services. However, because wheelchair users are elderly and disabled people, an existing user interface or operation method is hard to learn. Future work is focused on exploring interactive method of user interface on smartphone, making the user interface more user-friendly.

Acknowledgments The authors would like to thank the Science and Technology of China, Wuhan University of Technology for sponsoring the research activities under the grants of National Key Technologies Research and Development Program of China (2012BAJ05B07), International cooperation projects of Hubei province (2011BFA012), The Fundamental Research Funds for the Central Universities (2013-JL-005), and the joint bilateral Italy/China project "Smart Personal Mobility

Systems for Human Disabilities in Future Smart Cities" (N. CN13MO7). This work has been carried out by the research groups of the "Joint Lab. of Internet of Things" at Wuhan University of Technology and University of Calabria.

References

1. Estrin, D., Sim, I.: Open mHealth architecture: an engine for health care innovation. Science **330**(6005), 759–760 (2010)
2. Campbell, A.T., Eisenman, S.B., Lane, N.D., Miluzzo, E., Peterson, R.A., Lu, H., Ahn, G.S.: The rise of people-centric sensing. IEEE Int. Comput. **12**(4), 12–21 (2008)
3. Jara, A.J., Lopez, P., Fernandez, D., Castillo, J. F., Zamora, M. A., Skarmeta, A.F.: Mobile digcovery: discovering and interacting with the world through the Internet of things. Pers. Ubiquitous Comput. **18**(2), 323–338 (2014)
4. Fortino, G., Guerrieri, A., Lacopo, M., Lucia, M., Russo, W.: An agent-based middleware for cooperating smart objects. In Highlights on Practical Applications of Agents and Multi-Agent Systems. pp. 387–398. Springer, Heidelberg (2013)
5. Fortino, G., Guerrieri, A., Russo, W., Savaglio, C.: Middlewares for smart objects and smart environments: overview and comparison, in internet of things based on smart objects: technology, middleware and applications. Springer Series on the Internet of Things (2014)
6. Fortino, G., Guerrieri, A., O'Hare, G., Ruzzelli, A.: A exible building management framework based on wireless sensor and actuator networks. J. Netw. Comput. Appl. **35**, 1934–1952 (2012)
7. Bettini, C., Brdiczka, O., Henricksen, K., Indulska, J., Nicklas, D., Ranganathan, A., Riboni, D.: A survey of context modelling and reasoning techniques. Pervasive Mob. Comput. **6**(2), 161–180 (2010)
8. Rachuri, K.K., Efstratiou, C., Leontiadis, I., Mascolo, C., Rentfrow, P.J.: METIS: Exploring mobile phone sensing offloading for efficiently supporting social sensing applications.In Pervasive Comput. Commun. (PerCom), 2013 IEEE International Conference on, 85–93 March 2013
9. Li, B., Li, W. F.: Research on integration of WSN and RFID technology [J]. Comput. Eng. **9**(048), 127–129 (2008)
10. Fortino, G., Giannantonio, R., Gravina, R., Kuryloski, P., Jafari, R.: Enabling effective programming and flexible management of efficient body sensor network applications. IEEE Trans. Hum. Mach. Syst. **43**(1), 115–133 (2013)
11. Raveendranathan, N., Galzarano, S., Loseu, V., Gravina, R., Giannantonio, R., Sgroi, M., Jafari, R., Fortino, G.: From modeling to implementation of virtual sensors in body sensor networks. IEEE Sens. J. **12**(3), 583–593 (2012)
12. Bellifemine, F., Fortino, G., Giannantonio, R., Gravina, R., Guerrieri, A., Sgroi, M.: SPINE: A domain-specific framework for rapid prototyping of WBSN applications. Softw. Pract. Experience, Wiley **41**(3), 237–265 (2011). doi:10.1002/spe
13. Li, W., Bao, J., Shen, W.: Collaborative wireless sensor networks: A survey. In Systems, Man, and Cybernetics (SMC), 2011 IEEE International Conference on, 2614–2619 October 2011
14. Fortino, G., Guerrieri, A., Russo, W.: Agent-oriented smart objects development. In Computer Supported Cooperative Work in Design (CSCWD), 2012 IEEE 16th International Conference on, 907–912 May 2012
15. Shen, W., Xue, Y., Hao, Q., Xue, H., Yang, F.: A service-oriented system integration framework for community-based independent living spaces. IEEE International Conference on Systems, Man, and Cybernetics (SMC) 2626–2631 (2011)
16. Simpson, R.C.: Smart wheelchairs: a literature review. J. Rehabil. Res. Dev. **42**(4), 423–438 (2005)
17. Simpson, R.C., LoPresti, E., Cooper, R.A.: How many people would benefit from a smart wheelchair? J. Rehabil. Res. Dev. **45**(1), 53–72 (2008)

18. Satoh, Y., Sakaue, K.: An omnidirectional stereo vision-based smart wheelchair. EURASIP J. Image and Video Process. **2007**(1), 087646, 1–11, Chicago (2007)
19. Matsumoto, O., Komoriya, K., Hatase, T., Yuki, T., Goto, S.: Intelligent Wheelchair Robot TAO Aicle", Service robot applications, Chap. 4, Takahashi, Y. (Ed.), 55–70 (2008)
20. Bonci, A., Longhi, S., Monteriù, A., Vaccarini, M.: Navigation system for a smart wheelchair. J. Zhejiang Univ. Sci. **6A**(2), 110–117 (2005)
21. Parikh, S.P., Valdir Grassi JR., Kumar, V., Jun Okamoto JR.: Integrating human inputs with autonomous behaviors on an intelligent wheelchair platform. IEEE Int. Syst. **22**(2), 33–41 (2007)
22. Kuo, C.-H.: Mobility Assistive Robots for Disabled Patients. Service robot applications, Chap. 3, Takahashi, Y. (Ed.), 37–54 (2008)
23. Postolache, O.A., Silva Girao, P.M.B., Mendes, J., Pinheiro, E.C., Postolache, G.: Physiological parameters measurement based on wheelchair embedded sensors and advanced signal processing. IEEE Trans. Instrum. Measur. **59**(10), 2564–2574 (2010)

Experiments with a Sensing Platform for High Visibility of the Data Center

João Loureiro, Nuno Pereira, Pedro Santos and Eduardo Tovar

Abstract Data centers are large energy consumers and a substantial portion of this power consumption is due to the control of physical parameters, which bring the need of high efficiency environmental control systems. In this work, we describe a hardware sensing platform specifically tailored to collect physical parameters (temperature, pressure, humidity and power consumption) in large data centers. Our system architecture is composed of Smart Objects, the datacenter racks, that cooperate to contribute for the overall goal of finding opportunities to optimize energy consumption and achieving energy-efficient data centers. We also introduce an analysis of the delay to obtain the sensing data from the sensor network. This analysis provides an insight into the time scales supported by our platform, and also allows to study the delay for different data center topologies. Finally, we exemplify some capabilities of the system with a real deployment.

1 Introduction

Data center's large power consumption justifies a special attention to the design of energy efficient data centers. Power usage effectiveness (PUE) has become the metric

João Loureiro is supported by the government of Brazil through CNPq, Conselho Nacional de Desenvolvimento Científico e Tecnológico.

J. Loureiro (✉) · N. Pereira · P. Santos · E. Tovar
CISTER/INESC-TEC, ISEP, Polytechnic Institute of Porto, Porto, Portugal
e-mail: joflo@isep.ipp.pt

N. Pereira
e-mail: nap@isep.ipp.pt

P. Santos
e-mail: pjsol@isep.ipp.pt

E. Tovar
e-mail: emt@isep.ipp.pt

G. Fortino and P. Trunfio (eds.), *Internet of Things Based on Smart Objects*,
Internet of Things, DOI: 10.1007/978-3-319-00491-4_10,

to measure data center efficiency. It measures how much of the total energy consumed is really spent on IT work other than on facility's overhead, like lightning, cooling and power distribution, and it is given by: PUE = (IT Equipment Energy + Facility Overhead) / Energy IT Equipment Energy. It is desirable to measure it with a high spatial and temporal granularity, so that the PUE metric is as accurate as possible and to enable better understanding of the power consumption distribution in the data center. This better understanding may lead to great reductions through e.g. better load balancing, power distribution, or reduced air conditioning usage [1].

To have a full picture of the data center environment, it is important to collect air pressure, temperature, humidity and power consumption data at a high granularity (in time and space). The relevance of collecting these parameters is discussed in the next paragraphs.

In a typical data center, IT equipment is organized into rows, with a cold aisle in front, where cold air enters the equipment racks, and a hot aisle in back, where hot air is exhausted. Computer-Room Air Conditioners (CRACs) are commissioned to cycle the air, by pushing the cold air and returning the hot air to be cooled again. The CRAC systems are responsible for a big share of the facility overhead energy, and in order to achieve a more uniform thermal profile, special effort must be given on airflow distribution, by preventing cold and hot air from mixing and by eliminating any hot-spots. Better understanding of the airflow can be addressed by placing pressure and temperature sensors.

By measuring the local pressure, it is possible to estimate the speed and direction of the airflow between the sensed points and possibly identify unwanted mixtures or flow bottlenecks, as shown in [2]. It can also be used for workload-balancing among servers like in [3], where the patented application describes a system that uses a load balancer to shift tasks among servers based on their particular cooling needs, which is related to air pressure drop across the server. With fine grained temperature measurement it becomes easy to localize hot-spots, and by crossing this with pressure data, a better picture of the airflow can be taken, leading to better tuned CRAC systems.

Another important environmental parameter is the local humidity. Higher relative humidity decreases the chances of static electrical discharges that can damage the IT equipment and, at the same time, increases the heat transfer from the server to the cooling airflow. But too much water particles in the air reduces the lifetime of the IT equipment and increases the chance of water condensation at the cold aisles, which is not desirable. Several entities, such as the American Society of Heating, Refrigerating & Air-Conditioning Engineers (ASHRAE), provide guidelines with allowed and recommended values of relative humidity, as well as for dry bulb temperature, maximum dew point, maximum elevation and maximum rate of temperature changes, as seen in [4].

We present a sensing platform for collection of temperature, pressure, humidity and local power consumption (at rack or even server level). The development of the platform was centered on the specific application scenario of energy optimization in large data centers, focusing on high resolution sensing: several sensing points per rack, sampled at sub-seconds time intervals. Evidently, for such system to be practical, cost is an important factor to consider.

Our system architecture is composed of Smart Objects (SO), i.e. autonomous physical objects that also include sensing/actuation, processing, storage and networking capabilities [5, 6]. These SO cooperate to achieve a common goal, in our case, each rack (SO) provides access to processed data from its embedded sensor network, as a contribution to the overall goal of achieving high energy efficient IT rooms and data centers.

The middleware has essential role for this goal. As mentioned in [7], it provides general and specific abstractions, to allow building up complete software structures. Some generic midleware infrastructures were proposed, as in [8] for example. Ours was suited to the data-center context, more specifically to provide means for data-logging, visualization tools, alarm monitors, feedback data for the CRAC systems, or any other application that can be developed in the future. We addressed the midleware solution developed in our previous work at [9], which is not in the scope of this work.

In this chapter we will detail the design of the sensor network platform and develop an analysis of the time to obtain the sensing data from the nodes. This is done in order to study the time scales supported by our platform, and also allows to study the delay for different data center topologies. We also exemplify some capabilities of the system with a real deployment.

2 Related Work

Green data centers have received considerable attention in recent research literature. Some recent approaches rely on building software models through a joint coordination of cooling and load management [10, 11], or by formulating an energy minimization problem, subject to service delay and Quality of Service (QoS) constraints. In this class it is worth to mention dynamic voltage scaling [12, 13] and on/off power management schemes [14–16]. The complexity of data center airflow and heat transfer is compounded by each data center facility having its own unique layout, so achieving a general model is difficult [17]. For example, in [10], authors stress that their model has several parameters that need to be determined for specific applications.

Given such models, acquiring real-time data at a fine enough spatial and temporal resolution becomes an important topic, as this data can be used to validate models and keep their inputs updated at run-time. Nevertheless, this problem poses new challenges and research issues concerning the type, number and placement of sensors [17].

Some works [18, 19] pushed in the direction of deploying wireless sensor nodes and monitor the thermal distribution, to figure out how to avoid hot-spots and overheating conditions. We differ from such approaches in the sense that we want very fine-grained (in space and time) gathering of power and environmental parameters, including physical quantities other than temperature. Using a mixed wire/wireless solution, [18] obtained a average one-round collection time of approximately 6 s for 50 nodes. They also deployed 694 sensor nodes in a data-center, reading every cluster of 4 at most at every 30 s. In this work [18], for every cluster there was a wireless station and nodes where powered via USB, which makes the system dependent on having a powered USB port available (this might be a problem, since the server to where the node is connected to cannot be powered off, for example). A pure wireless solution was presented in [19], where it was reported a deployment of 107 battery powered wireless nodes, taking 3 s to sample all of them (not considering data losses). The experiment only lasted for 35 days before the battery had to be replaced, which is not practical for large, long-lived deployments.

Our proposed system is based on a hierarchical, modular, flexible and fine-grained sensor network architecture, where data is collected from heterogeneous sensors (including power), placed in each rack. The analysis of their inter-correlations will enable closer examination and a better understanding of the flow and temperature dynamics within each data center [20]. To our knowledge, no previous work enables correlating power and environment characteristics on a per rack or per-server granularity with such temporal resolution.

Multiple long-wavelength infrared image sensors can be used to capture thermal maps of an environment [21]. While thermal cameras are an interesting approach, we find that they suffer from several practical issues: (i) the current cost of thermal cameras is substantial, and, due to field-of-view limitations (data centers are typically organized in narrow rows), a high number of them should be required to cover a data center; (ii) mapping the view of the camera with the infrastructure being monitored is more challenging than relying on point sensors, and it is especially difficult to manage when changes are made to the layout of the data center (e.g., addition/removal of servers and racks), and (iii) by using cameras, the quantitative data analysis would need to be provided by computer vision, which is feasible, but requires a very specific tuning for each scenario and equipment. However, as claimed also by authors in [22], our system has provisions to support thermal image sensors as a smart sensor that can provide temperature field readings with a configurable resolution.

Another approach commonly used is to make measurements throughout the data center manually, or using mobile robots to automate this task [23]. This approach does not enable practical high-resolution real-time monitoring of the data center as our system does.

3 Overview

The proposed sensor network architecture is a combination of wired and wireless technologies, designed to achieve high spatio-temporal resolution of data center rooms, keeping system's flexibility and modularity, with a low latency and low cost.

Our system is designed to cover the data center first by a short range bus that covers the communication needs inside each rack, a longer range bus that covers each row in the data center and then wireless communication is used to gather the

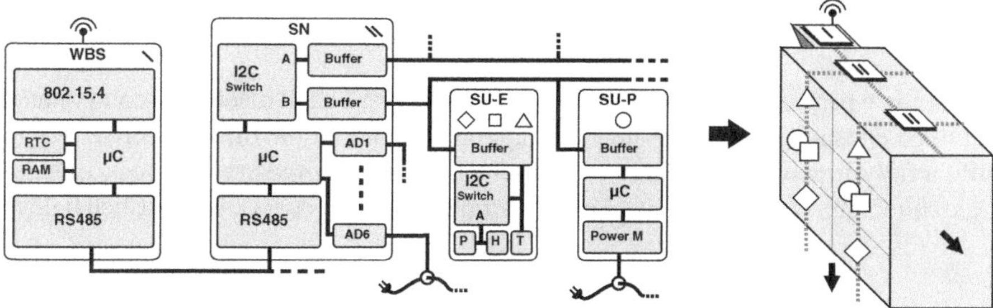

Fig. 1 Network architecture and layout

data from the entire data center room. Four different types of devices cover each of these levels (rack, row and room): (i) *Sensing Units* sense the physical parameters (temperature, pressure, humidity, and power) in each rack, then (ii) *Sensor Nodes* collect the sensing data for the entire rack, and (iii) Wireless Base Stations (*WBS*s), collect data from several Sensor Nodes in a row, as represented in Fig. 1. Finally, (iv) *Gateways* collect data from all of the *WBS*s in a data center room.

Starting at the lower level, our sensor network consists of two different types of Sensing Units: (a) a small passive sensing unit for measuring environmental quantities, with at most one temperature, one humidity and one pressure digital sensor, and (b) a power metering unit with real, active, and reactive power measurement capabilities, as presented in Fig. 1 by *SU-E* and *SU-P* respectively. The environmental Sensing Units can be manufactured according to the sensing and cost needs, by having any combination of sensors on it, what is represented by the three different shapes. Both sensing units deliver data to the next level in the hierarchy, through a wired short range bus (I2C), projected to cover only one rack of servers (back and front).

At the next level, the Sensor Node is responsible to collect the data of all the Sensing Units attached to it and possibly to perform simple data aggregation and sensor fusion before delivering it to the next level in the hierarchy using a longer range wired bus (MODBUS).

WBSs are responsible for querying the Sensor Nodes within their respective cluster, and again perform data aggregation, sensor fusion and data analysis. They communicate then with devices at the next level in the hierarchy to deliver the relevant data. Gateways then provide the data gathered from the sensor network to the data distribution system in a standard format. From this point on, sensing data is published at a publish/subscribe middleware that distributes the acquired data to different applications, where each of them will use such information with different proposes (alarms, data logging, visualization, etc).

Each Sensor Node can be connected up to 52 temperature sensors, 54 power meters, 14 pressure sensors and 14 humidity sensors. The following section describes in more detail each of the system components.

4 Platform Details

Well-known protocols, network architectures and of-the-shelf electronic components had to be chosen to compose the system, considering that the final objective was to build a fully functional, industry ready, sensor network with very low cost. Besides the architecture, the technology chosen to implement the network is described below.

4.1 Sensing Unit

With the popularization of two-wire I2C buses on motherboards, cellphones and on general embedded systems, many companies are nowadays developing sensors with digital I2C output, by embedding the micro-mechanical sensor, signal amplifiers, analogue to digital converters, memory and a I2C front end to manage with the communication on the bus. These Systems-on-Chip enable high accuracy and reliability measurements, since this decreases the probability of data corruption due to any external interference. It also prevents calibration issues found on pure-analogue sensors measurements, since digital sensors are factory calibrated and digitally compensated. Due to these reasons, I2C sensors were used to connect the several sensing units.

Some limitations of I2C buses had to be overcome to make its usage practical in this application. First, buffers had to be added as an interface between the I2C bus lines and every circuit board attached to it, in order to allow the I2C to operate over longer distances, by increasing the robustness of the logic signals of the standard I2C buses. Second, switches had to be added to every Sensing Unit on the bus in order to allow the usage of more than one sensor with non-configurable addresses, making it accessible from the main bus.

Figure 2b depicts one Sensing Units with temperature, humidity and pressure sensors. The temperature sensor used is a low cost and low power device with 1.5 °C accuracy, maximum resolution of 0.0625 °C and minimum and maximum conversion times between 27.5 and 300 ms. The humidity sensor has 1.8 %RH accuracy, with maximum 0.04 %RH resolution and minimum and maximum conversion times between 3 and 29 ms, both the temperature and humidity sensors suitable for the application, where the focus are in changes in major scales according to the ASHARE guidelines [4], which specifies a range of dew points between 5.5 °C (for 60 %RH) and 15°C. The pressure sensor ranges from 300 to 1100 hPa, with an accuracy of +-1 hPa typical and 0.03 hPa of resolution with minimum and maximum conversion times between 3 and 25.5 ms, also suitable for the application, where typical pressure variation values inside data center's are in greater orders of magnitude, as seen in [2].

The Power Meter Sensing Unit is composed by a dedicated chip which interfaces with the power line, and provide real, reactive, and apparent power measurements to the embedded computational unit, which is responsible for interfacing with the I2C bus as a slave, and to deliver such information to the master, at the next level.

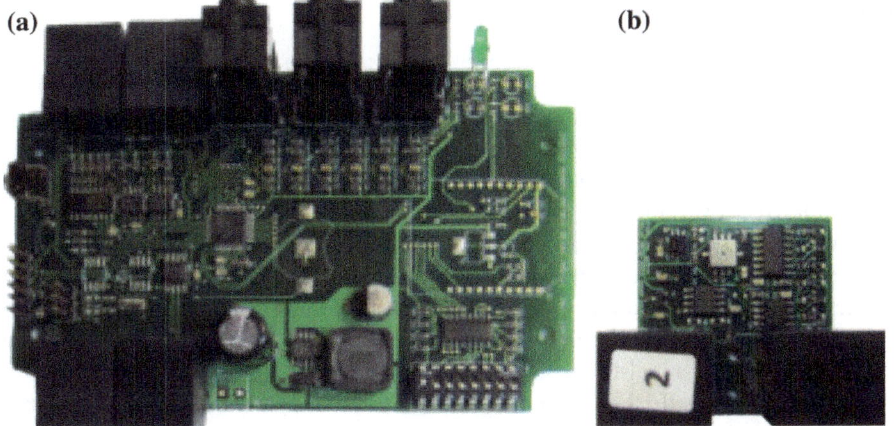

Fig. 2 Hardware platforms. **a** Sensor node, **b** Sensing unit

To both Sensing Units, the power is carried into the same cable as the I2C data, and locally converted from 5 to 3.3 V by a low-drop LDO converter, for more stable and lower ripple power supply for the sensors, which are sensitive to such variations.

4.2 Sensor Nodes

A Sensor Node is a communication/computation enabled device, physically linked over the I2C bus (also trough buffers) to a number of Sensing Units. The Sensor Nodes gather the data from the Sensing Units and, in turn, answer to data requests from the WBS. Figure 2a depicts a Sensor Node.

To keep cost and complexity low at this tier of the network architecture, the Sensor Nodes communicate with one Wireless Base Station (WBS) over a bus, e.g., using a RS485/MODBUS technology [24]. In particular, the WBS node acts as a local coordinator and master of the bus.

The Sensor Node is also composed by: (i) six analogue inputs suited for current measurement, connected to external current transducers attached to the power lines, as a cheap and simple alternative for basic current measurement; (ii) two I2C buffered ports through one switch, responsible for duplicating the bus capacity in terms of addressable devices, and enabling a better mechanical placement for cables to go to the back and front of a rack, and (iii) one RS485 port for the MODBUS.

The power supply for the Sensor Nodes is carried by a twisted pair cable, along with the MODBUS data, in another pair. At every Sensor Node, a high efficiency DC-DC step down converter, converts from 48 to 5 V for the local supply. This is an important feature as it reduces the number of cables that connect to each node, facilitating installation of the devices.

4.3 Wireless Base Stations (WBSs)

The WBS is directly connected to a power source and supplies power through a twisted pair cable to all the Sensor Nodes in that bus. In all the nodes on this bus, the voltage is locally converted to lower values by a step-down switched power supply for a higher system efficiency. Wires running in the same cable form a serial data bus (MODBUS over a RS485 connection) that interconnects the Sensor Nodes.

The WBS is based on the same printed circuit board as the Sensor Node, missing the sensors interfaces, and with some extra components, like one external non-volatile ferrite random access memory (FRAM), used as a buffer and for diagnosing the system in cases of failures or power cuts (by keeping the last operational state). The WBS also includes a real-time clock used for time stamping the data packets.

The WBSs act as IEEE 802.15.4 cluster heads and are connected with each other in a mesh topology. A common Gateway is in charge of gathering measurements and sending them over long range communication technology (e.g., WiFi, Ethernet). In terms of HW platforms, the WBS node will be the same platform as a generic Sensor Node, with an on-board ZigBee radio. Thus, each Sensor Node can become a WBS with minimal modifications, i.e., just by plugging the wireless module and uploading a different firmware.

4.4 Gateways

The sensor network can have one or more Gateways. Gateways maintain representations of the data flows from the sensor network to the data distribution system. They perform the necessary adaptation of the data received from the WSN. The gateways can be deployed as one per room serving all the rows of racks in that room; more gateways can also be deployed to improve radio coverage, for load-balancing or for redundancy.

5 Delay Analysis

When performing deployments of our system, we need to answer questions related to how the network should be deployed (for example, we can choose how many sensing points should we deploy per WBS) and what is the impact of this in the performance of the network. To answer such questions we have developed an analysis of the time to transmit sensor data. This analysis also shows that our system can exhibit very low delays in the presence of a large number of sensing points.

This analysis enables us to study the communication delay as we add Sensor Nodes to the network. We consider that each Sensor Node added has N_{su-sn} Sensing Units attached to it, where each Sensing unit has three 16 bit sensors. For every

N_{sn-wbs} Sensor Nodes added to the network, one WBS has to be added also. The total number of Sensor Nodes is defined as N_{sn}. Clearly, these parameters (N_{su-sn} and N_{sn-wbs}) are defined according to the topology of the deployment and of the data center room.

5.1 Calculating the Response Time

The response time R required to collect data from all the sensors is given by adding together the time to transmit all the wireless requests to all WBS (t_{req}) and also the corresponding replies (t_{rep}), as given by Eq. (1).

$$R = \left(t_{req} + t_{rep}\right) \tag{1}$$

The time to transmit all requests is computed by the sum of the time required to transmit a request to each WBS (there are $\lceil \frac{N_{sn}}{N_{sn-wbs}} \rceil$ WBS:s in the network) with the worst-case blocking time, B_{mb}, is given by Eq. (2).

$$t_{req} = \left\lceil \frac{N_{sn}}{N_{sn-wbs}} \right\rceil \times \left(t_{wtx}(S_{wreq}) + B_{mb}\right) \tag{2}$$

where the $t_{wtx}(S_{wreq})$ is the time to transmit a request packet in the wireless 802.15.4 network including all protocol overhead for a packet with S_{wreq} bits of payload, and will be defined later. B_{mb} is a constant given by the longest data transaction over the MODBUS, which corresponds to the largest task to be executed by the WBS in a non preemptive system.

The time to transmit all replies is given by Eq. (3) as follows:

$$t_{rep} = \left(\left\lfloor (N_{su-sn} \times N_{sn}) \times \frac{S_{sd}}{S_{mwp}} \right\rfloor + 1\right) \times t_{wtx}(S_{mwp}) \tag{3}$$

where S_{sd} is the size of the sensor data to be transmitted by each Sensor Unit and S_{mwp} is maximum wireless data payload, after accounting for all protocols headers. $t_{wtx}(S_{mwp})$ is the time to transmit a packet in the wireless IEEE 802.15.4 network with the maximum possible payload (mwp bits) and will be defined in Sect. 5.2.

5.2 Calculating the Wireless Transmission Time

The reasoning applied to calculating the wireless transmission time ($t_{wtx}(S)$) is similar to the one found in [25, 26] when analyzing the maximum theoretical throughput of a non-beacon enabled IEEE 802.15.4. The time to send a IEEE 802.15.4 packet with payload size of S bits if given by:

$$t_{wtx}(S) = T_{ib} + t_{ppdu}(S) + T_{ack} + T_{ifs} \quad (4)$$

where T_{ib} is the initial back-off period, which depends on the parameter $macMinBE$, and, by default, $macMinBE = 3$, resulting in $T_{ib} = 1120\ \mu s$). The time to transmit the PHY protocol data unit (ppdu) with a payload size of S bits is denoted by $t_{ppdu}(S)$. The time to transmit an acknowledgment is defined as $T_{ack} = T_{ackppdu} + T_{rxtx} = 544\ \mu s$ since it must include the time to send the acknowledgment packet ($T_{ackppdu} = 352\ \mu s$ as defined in the standard [27]) and the time for the transceiver to switch from receive to transmit ($T_{rxtx} = 192\ \mu s$ is the maximum value defined in [27], and this is the value found in the 802.15.4 transceivers employed [28]). The inter-frame spacing (IFS), T_{ifs}, is set to the value of the long IFS defined by the standard, 640 μs (actually, this is only used when the size of the MAC protocol data unit (MPDU) to be sent is above or equal to 18 bytes [27]).

The time to transmit the ppdu with a payload of size S bits, can be defined as:

$$T_{ppdu}(S) = (S_{hdr} + S_{zbee} + S + S_{ftr}) \times \tau_{bit} \quad (5)$$

where S_{hdr} is the sum of the sizes of the synchronization header (SHR), PHY header (PHR) and MAC header (MHR; from [27]: $S_{SHR} = 40$; $S_{PHR} = 8$; $S_{MHR} = 56$ bits). The size of the ZigBee protocol headers is $S_{zbee} = 41 * 8$ bits, and the size of the MAC footer is $S_{ftr} = 16$ bits. The time to transmit one bit is $\tau_{bit} = 4\ \mu s$ (for a data rate of 250 kbps).

5.3 Delay Results

Instantiating the response time given by Eq. (1) results in Fig. 3a and b. For these calculations, we have used $S_{wreq} = 16$ bits (a request with a two-byte identifier) and $S_{mwp} = 576$ bits (the maximum IEEE 802.15.4 payload minus the overhead defined in Eq. (5)).

With Fig. 3a, we analyzed the impact of adding *SN* to the network with varying N_{su-sn}. As expected, the increase in the delay is linearly proportional to the N_{su} on the network, when keeping N_{sn-wbs} constant. The higher the N_{su-sn}, higher is the slope. This is expected because the amount data is constantly added as we added *SU*, however there is a more pronounced increase in response time whenever a *WBS* is added. In this case, at every 20 $SN's$ added, a higher step is expected due to the overhead of adding wireless links to the network.

Figure 3b now shows the case where N_{su-sn} is fixed, and we vary N_{sn} over N_{sn-wbs}. With smaller N_{sn-wbs}, the response time increases very pronouncedly. For example, if there is only one *SN* per *WBS*, for every *SN* added to the network, one more wireless link will be added, causing significant increase in the response time. By increasing N_{sn-wbs}, this effect decreases very rapidly also.

Figures 4a and b present another aspect related to the network topology, which must be considered when designing the network. The horizontal line in both plots

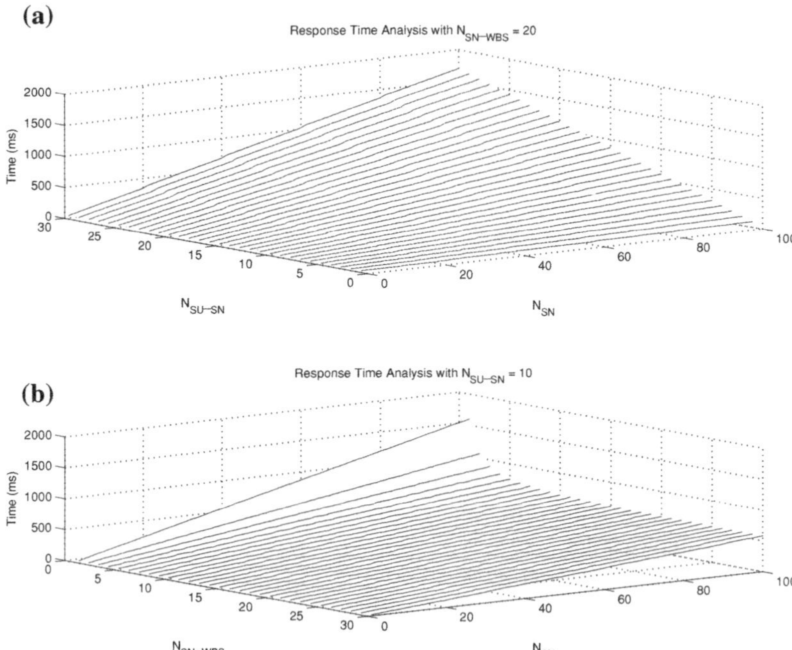

Fig. 3 Network response time for different possible configuration scenarios. **a** Response time analysis with $N_{sn-wbs} = 20$, **b** response time analysis with $N_{su-sn} = 10$

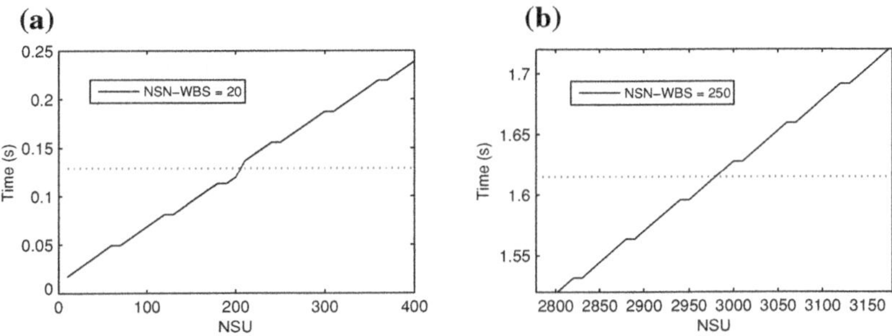

Fig. 4 Network response time. **a** $N_{sn-wbs} = 20$, **b** $N_{su-wbs} = 250$

shows the time to gather the data from all Sensor Nodes attached to the WSB (20 Sensor Nodes in Fig. 4a, and 250 in Fig. 4b). The way the network is designed, if one implements a network with N_{sn} below the intersection between the horizontal line and the response time, the wireless communication cycle of the WBS will be faster than the communication cycle on the MODBUS. Thus, the WBS would repeatedly transmit data from previous communication cycles. N_{sn-wbs} should be set such that the lines intersect at the desired N_{sn}. Something that can be easily found, given the analysis presented in this section. In Fig. 4a and b we can see a stepped behavior of the response time, with the growth of the N_{su}. One step happens at each $6 \times N_{su-sn}$. The reason for this step is that, as we add Sensor Nodes, there is the need for and extra packet to be sent (the length of the packet and number of packets needed depends on N_{su-sn} and also on the maximum payload *mwp*). In this scenario, the sensor data for the 7th Sensor Nodes fits in the same number of packets, and thus the delay does not increase. A bigger step is given at every N_{sn-wbs}, due to the overhead of adding one WBS more.

6 Data Center Visualization

The deployment of the described system in the data center enables interesting opportunities to have better insight into the data center conditions. In this section we will briefly provide some data from a real deployment. This data was selected for its relevance in showing different aspects of the data center conditions that are enabled by the deployment.

The deployment in this section was performed in a data center room owned by the largest telecommunications operator in Portugal. All racks were fitted with two temperature sensors in the front and two temperature sensors in the back. Per row, sensors with additional humidity and pressure sensors were deployed such that the row had three racks (at the top, end, and middle of the row) with such sensors.

Previously, data center operators add a few options to gather such pictures of the data center conditions (e.g. thermal cameras or mobile robots), as discussed in the Related Work Section. We claim that our systems enables high-resolution and real-time monitoring of the data center. Something not available in practical systems to date. Our system enables real-time maps temperature, pressure and humidity. These maps are useful to have a detailed picture of the data center conditions. Because the information is collected in real-time, automated control of the data center physical conditions can be enabled.

6.1 Real-Time Thermal Profiling

To illustrate the maps enabled by our tool and to better demonstrate and exploit the capabilities and improvements that our tool can bring the data center management, we have chosen to depict the thermal map of one representative row, as shown in Fig. 5.

By analysing Fig. 5, it is possible to see the cold air concentration at the bottom of the racks. It happens because the cold air comes from the perforated tiles on the floor, and due to its higher density, compared to the hot air, it stays at the floor level. Enough pressure drop from the bottom to the top of the racks would be required in order to guarantee the cold air flow till the air intake of the server on the top of the racks.

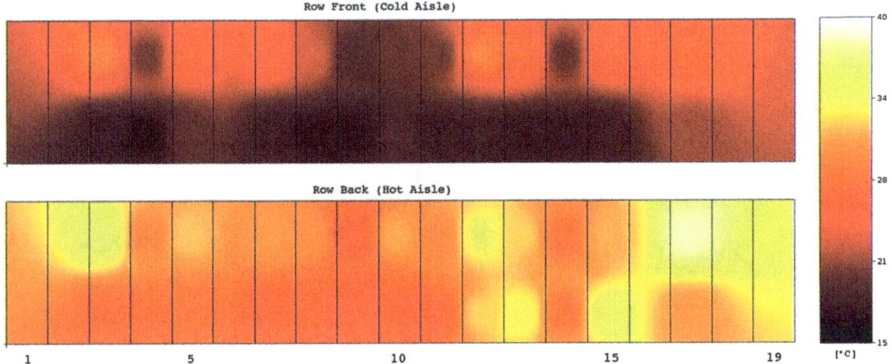

Fig. 5 Heat map of one data center row

In more detail Fig. 5, shows that on the top region, the intake temperature is around 26 °C, which is much higher than the cooled air temperature, which is around 17 °C. This is also explained by some air recirculation of hot air from the hot aisle (with higher pressure) to the cold aisle. Therefore, the upper servers receive a mixture of hot and cold air, with an intermediate temperature between its output and the cold air temperature.

It is also possible to notice how the temperature at the bottom of the racks gradually rises on the last four racks of the row (to the right). Correlating this with the pressure data, it is possible to notice that the pressure drop between both extremes is not equally distributed, explaining why the cold air does not reach well the last four racks of the row.

Regarding the back side of the row, at the hot aisle, we can clearly see how the air output temperature is correlated with the input air temperature. The colder the air at the input, colder the output flow. Despite this, the workload can also significantly interfere on the heat transfer to the air. One example can be observed at racks 12 and 13, that, even having low temperatures at the air intake at the bottom of the rack, the output temperature raised much more, compared with the neighboring racks. This is a very common effect in heterogeneous data centers, harboring different types of machines, with different powers. Different heat outputs can also be found when workload moves between machines, for example due to workload management in a virtualized infrastructure.

6.1.1 Modifications and Discussion

An intervention was made to the row displayed in Fig. 5 in order to improve the temperature distribution. The intervention consisted in manually adjusting the perforated tiles located in the cold aisle of the row. Figure 6, presents the average temperature of some selected racks in the row, and allows us to see the evolution of the temperature

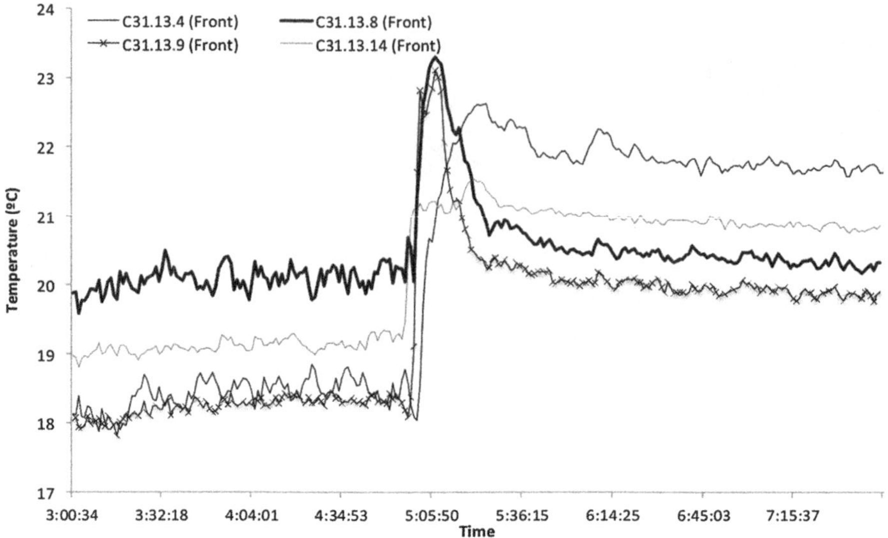

Fig. 6 Temperature of a row during perforated tiles adjustments. **a** Row front (Cold Aisle)

during that intervention, which took place around 5PM. We can see that the adjustments momentarily caused the temperature to rise, to then stabilize to a value slightly higher than before the intervention. With this intervention, the temperature distribution in the row became much more homogeneous, and the set point of air conditioning system could be increased, bringing direct savings. It is important to note that the data presented in Fig. 6 are provided directly by the monitoring tool developed, and facilitates the observation of how the data center conditions evolve throughout time.

All of the described issues found in this scenario are commonly found in data centers all over the world. In this case, small changes have brought a more homogeneous heat distribution. A second step, with a minor investment, could be to install curtains on the top of the racks, and on its extremity. It could prevent the mixture of cold and hot air, leading to a even better heat distribution and specially, minimizing the waste of energy due to the mixtures. The servers on the top would be directly impacted by this change, by receiving colder air on its intake, contributing to a better distribution. The PUE of the data center can be dramatically decreased with such simple actions.

Similar heat maps can be also obtained from the entire room with its view from the top, possibly presenting any of the different sensors types and position, grouped in any way. Even more, a 3D representation could be done, by using heat transfer models to estimate the surroundings of the measured point with improved accuracy. This models could include input parameters like rack or server instant power consumption, and temperature, humidity and pressure at the air input and output. Externally, this data could be correlated with the instantaneous workload of every server, allowing work reallocation to minimize hot spots, for example.

With such representations in real-time, the data center manager can have a new representation of the operational condition of it. Local actuation can be done without the need of data exchanges between higher and lower levels of the network, by having a *SN* locally acting on an automated perforated tile.

These actions could be supervised by *WBS′s* due to its better overall picture of the row. It could have some influence over the individual control parameters on every *SN* for a more general control.

6.2 Real-Time Power Profiling

The system developed can also be used to collect real-time data about the power being consumed in the data center. Power data can be collected by power circuit in each rack, or even per server, when the cost is justified.

We took advantage of this feature see if the workload a typical modern server influences the physical parameters measured by our system in a significant way and if this change can happen fast (with, for example, power). This can justify automated local actuation in the data center, and thus elicit the need for real-time data collection.

We deployed sensors around a rack server used for a virtualized infrastructure, in a way similar to a normal data center deployment. We have then measured how the temperature and current consumption varied in time and with changing server workload.

Figure 7a, b shows the power trace when the workload of the server changes almost instantaneously from an idle state to 100 % utilization. This change is reflected almost immediately in the power consumption as seen in the figure. This measurement incurs in the delay bounded by the Eq. (1). While much slower, we can see that the temperature also increases significantly as a result of this workload increase. However, it takes about 11 min to go up to the maximum of 44 °C.

Then, we have increased the utilization if steps of 25 % from 0 to 100 %. In Fig. 7b it is possible to see these steps reflected in the current consumed by the server. For the current, the first step presented a rise of 40 % over the background consumption, while the following steps rise 20 % approximately, showing that the power consumption and temperature have significant variations even for workloads much lower than 100 %.

To conclude, we verify that a physical parameter measured (power) does change very fast. The temperature, while being significantly slower still exhibits a large variation over time.

7 Conclusions

We have presented a platform for acquiring the physical parameters of a data center. This platform was developed as a mix of wired and wireless communicating nodes, such that it can enable flexible monitoring of the data center at a very high temporal

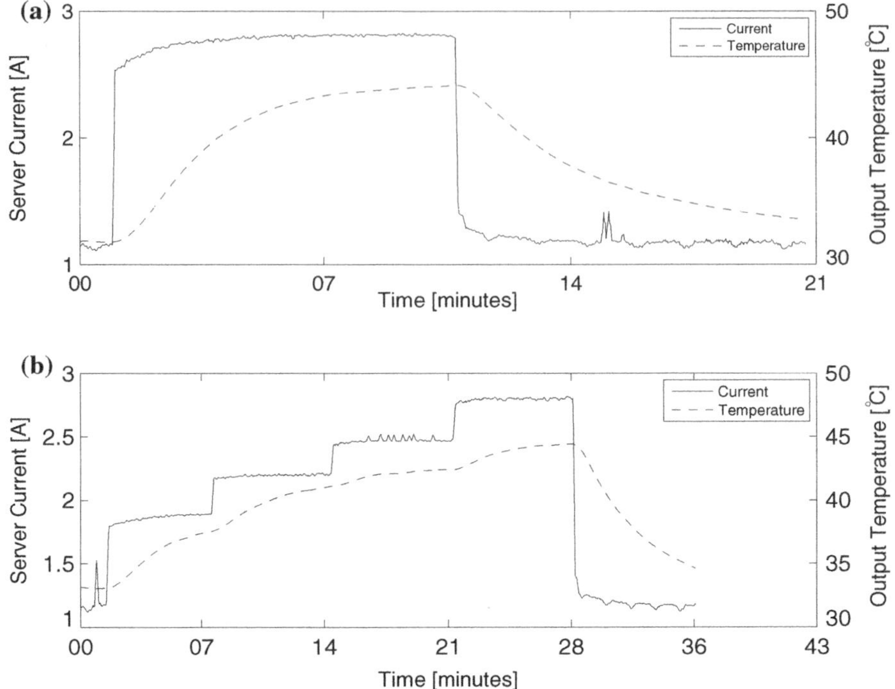

Fig. 7 Power and temperature traces. **a** Power and temperature trace in time, **b** power and temperature as utilization increases

and spatial resolution of the sensor measurements, while keeping the cost per sensing point very low. Compared to previous work, we enable much higher sensing resolution (several sensing points per rack, sampled at sub-second frequency), maintaining cost low and ease of installation.

We also presented an analysis of the delay of our system. This analysis enabled us to study the communication delay as we add Sensor Nodes to the network, and has shown that our system can exhibit very low delays in the presence of a large number of sensing points. This analysis also allows to try different network deployments and check the trade off between different topologies (described by parameters N_{su-sn} and N_{sn-wbs}) and the resulting delay.

Our experiments have exemplified the data that can be collected by the system and that the physical parameters measured by the system are impacted directly and in a dynamic way by the workload of the servers. Acquiring physical parameters at a very high resolution is important to find opportunities to optimize energy consumption, minimize local hot-spots, achieve more accurate predictive maintenance, perform more accurate billing, and it also enables very fast response to changes in the measured parameters, including automated actuation.

Acknowledgments This work was supported by National Funds through the FCT-MCTES (Portuguese Foundation for Science and Technology) and by ERDF (European Regional Development Fund) through COMPETE (Operational Programme 'The- matic Factors of Competitiveness'), within projects Ref. FCOMP-01-0124-FEDER-022701 (CISTER), FCOMP-01- 0124-FEDER-012988 (SENODs) and FCOMP-01-0124-FEDER-020312 (SMARTSKIN).

References

1. Google. Google's Green Data Centers : Network POP Case Study
2. Brey, T., Lembke, P., Prisco, J., Abbott, K., Cortese, D., Hazelrigg, K., Larson, J., Shaffer, S., North, T., Darby, T.: Case Study: The ROI of Cooling System Energy Efficiency Upgrades
3. Michael, A.M., Paleczny, M.: Load Balancing Tasks in a Data Center Based on Pressure Differential Needed for Cooling Servers (2012)
4. TC ASHRAE. 2011 thermal guidelines for data processing environments expanded data center classes and usage guidance. ASHRAE, pp 1–45 (2011)
5. Atzori, L., Iera, A., Morabito, G.: The internet of things: a survey. Comput. Netw. **54**(15), 2787–2805 (2010)
6. Fortino, G., Guerrieri, A., Lacopo, M., Lucia, M., Russo, W.: An agent-based middleware for cooperating smart objects. In: Highlights on Practical Applications of Agents and Multi-Agent Systems, pp. 387–398. Springer, New York (2013)
7. Fortino, G., Guerrieri, A., Russo, W., Savaglio, C.: Middlewares for smart objects and smart environments: overview and comparison. In: Internet of Things Based on Smart Objects, Technology, Middleware and Applications, Internet of Things. Springer. isbn: 978-3-319-00490-7 (2014)
8. Kawsar, F., Nakajima, T., Park, J.H., Yeo, S.-S.: Design and implementation of a framework for building distributed smart object systems. J. Supercomput. **54**(1), 4–28 (2010)
9. Pereira, N., Tenina, S., Tovar, E.: A microscope for the data center. In: Wireless Algorithms, Systems, and Applications, pp. 619–630, Springer (2012)
10. Parolini, L., Sinopoli, B., Krogh, B.H.: Reducing data center energy consumption via coordinated cooling and load management. In: Proceedings of the 2008 Conference on Power Aware Computing and Systems, HotPower'08, pp. 14–14, Berkeley, CA, USA (2008). USENIX Association
11. Zhou, R., Wang, Z., Bash, C.E., McReynolds, A.: Data center cooling management and analysis—a model based approach. In: 28 Annual Semiconductor Thermal Measurement, Modeling and Management Symposium (SEMI-THERM), San Jose, California, USA (2012)
12. Bohrer, P., Elnozahy, E.N., Keller, T., Kistler, M., Lefurgy, C., McDowell, C., Rajamony, R.: Chapter the case for power management in web servers, In: Melhem, R., Graybillpp, R. (eds.) Power aware computing. pp. 261–289. Kluwer Academic Publishers, Norwell (2002)
13. Tibor, H., Tarek, A., Kevin, S., Xue, L.: Dynamic voltage scaling in multitier web servers with end-to-end delay control. IEEE Trans. Comput. **56**(4), 444–458 (2007)
14. Xu, R., Zhu, D., Rusu, C., Melhem, R., Mossé D.: Energy-efficient policies for embedded clusters. In: Proceedings of the 2005 ACM SIGPLAN/SIGBED Conference on Languages, Compilers, and Tools for Embedded Systems, LCTES '05, pp. 1–10, New York, NY, USA (2005). ACM
15. Meisner, D., Gold, B.T., Wenisch, T.F.: Powernap: eliminating server idle power. In: Proceedings of the 14th International Conference on Architectural Support for Programming Languages and Operating Systems, ASPLOS '09, pp. 205–216, New York, NY, USA (2009) ACM
16. Wang, S., Chen, J.-J., Liu J., Liu, X.: Power saving design for servers under response time constraint. In: Proceedings of the 2010 22nd Euromicro Conference on Real-Time Systems, ECRTS '10, pp. 123–132, Washington, DC, USA (2010). IEEE Computer Society

17. Jeffrey, R., Yogendra, J.: Modeling of data center airflow and heat transfer: state of the art and future trends. Distrib. Parallel Dat. **21**(2–3), 193–225 (2007)
18. Liang C.-J.M., Liu, J., Luo, L., Terzis, A., Zhao F.: Racnet: a high-fidelity data center sensing network. In: Proceedings of the 7th ACM Conference on Embedded Networked Sensor Systems, SenSys '09, pp. 15–28, New York, NY, USA (2009). ACM
19. Weiss, B., Truong, H.L., Schott, W., Scherer, T., Lombriser, C., Chevillat, P.: Wireless sensor network for continuously monitoring temperatures in data centers. IBM RZ 3807 (2011)
20. Schmidt, R.R., Cruz, E.E., Iyengar, M.: Challenges of data center thermal management. IBM J. Res. Dev. **49**(4.5):709–723 (2005)
21. Karlsson, J.F., Moshfegh, B.: Investigation of indoor climate and power usage in a data center. Energ Buildings **37**(10), 1075–1083 (2005)
22. Viswanathan, H., Lee, E.K., Pompili, D.: Self-organizing sensing infrastructure for autonomic management of green datacenters. IEEE Netw. **25**(4):34–40 (2011)
23. Lenchner, J., Isci, C., Kephart, J.O., Mansley, C., Connell, J., McIntosh, S.: Towards data center self-diagnosis using a mobile robot. In: Towards data center self-diagnosis using a mobile robot, pp. 81–90 (2011). ACM
24. Modbus over serial line—specification & implementation guide—v1.0. http://www.modbus.org/docs/Modbus_over_serial_line_V1.pdf (2002)
25. Latré, B., De Mil P., Moerman, I., Dhoedt, B., Demeester, P., Van Dierdonck N.: Throughput and delay analysis of unslotted IEEE 802.15.4. JNW **1**(1):20–28 (2006)
26. Measuring effective capacity of IEEE 802.15.4 beaconless mode, volume 1, (2006)
27. IEEE. IEEE standard for information technology—telecommunications and information exchange between systems—local and metropolitan area networks—specific requirements - part 14.4: Wireless medium access control (MAC) and physical layer (PHY) specifications for low rate wireless personal area networks (LR-WPANs), October (2003)
28. Chipcon. CC2420 datasheet. http://www.chipcon.com/files/CC2420/Data///Sheet/1/3.pdf